THE
SEEDS
OF LIFE

THE
SEEDS
OF LIFE

From Aristotle to da Vinci, from
Sharks' Teeth to Frogs' Pants, the
Long and Strange Quest to Discover
Where Babies Come From

Edward Dolnick

BASIC
BOOKS
NEW YORK

Published by Basic Books, an imprint of Perseus Books, LLC, a subsidiary
of Hachette Book Group, Inc.

Books published by Basic Books are available at special discounts for bulk purchases
in the United States by corporations, institutions, and other organizations. For more
information, please contact the Special Markets Department at Perseus Books,
2300 Chestnut Street, Suite 200, Philadelphia, PA 19103, or call (800) 810-4145,
ext. 5000, or e-mail special.markets@perseusbooks.com.

Designed by Linda Mark

Library of Congress Cataloging-in-Publication Data
Names: Dolnick, Edward, 1952– author.
Title: Seeds of life : from Aristotle to Da Vinci, from sharks' teeth to frogs' pants, the
    long and strange quest to discover where babies come from / Edward Dolnick.
Description: New York : Basic Books, [2017] | Includes bibliographical references
    and index.
Identifiers: LCCN 2016054195| ISBN 9780465082957 (hardcover) |
    ISBN 9780465094967 (ebook)
Subjects: LCSH: Human reproduction—History. | Human reproduction—
    Mythology. | Human reproduction—Social aspects. | BISAC: SCIENCE / History. |
    HISTORY / Social History. | SCIENCE / Life Sciences / General.
Classification: LCC QP251 .D59 2017 | DDC 612.6—dc23 LC record available at
    https://lccn.loc.gov/2016054195

LSC

10 9 8 7 6 5 4 3 2 1

For Lynn, and Sam and Ben

Night and day, the ignorant as well as the learned give themselves over to the pleasure of making children. But no one knows how he has engendered his own progeny.

—Vittore Cardelini, Italian physician and author, 1628

CONTENTS

**PART FOUR: THE CLOCKWORK TOPPLES
AND A NEW THEORY RISES**

TIME LINE

1490—Leonardo da Vinci makes a cutaway drawing of a man and woman having sex.
1492—Columbus sets sail.
1543—Andreas Vesalius publishes one of the masterpieces in the history of anatomy.
1543—Copernicus says that the Earth goes around the sun, not vice versa.
1628—William Harvey shows that the heart is a pump.
1651—Harvey declares that "everything comes from the egg."
1669—Jan Swammerdam argues that God created all the generations of animals at the dawn of time, one inside the next like Russian dolls.
1672—Regnier de Graaf (almost) proves that female mammals have eggs.
1674—Antony van Leeuwenhoek sees countless "tiny animals," invisible to the naked eye, in a drop of pond water.
1677—Leeuwenhoek sees spermatozoa by the millions.
1694—Nicolaas Hartsoeker draws a miniature man inside a sperm cell.
1741—Abraham Trembley cuts a tiny organism called a hydra into pieces. Miraculously, each piece grows into a complete creature.
1745— French scientists propose a new theory of how living organisms develop: life is regulated not by clockwork but by a force akin to gravity.
1752—Ben Franklin flies a kite during a thunderstorm and proves that lightning is electrical.
1770s—Lazzaro Spallanzani puts male frogs in boxer shorts.
1776—American Revolution begins.
1791—Luigi Galvani zaps frog legs with electricity.
1818—Mary Shelley publishes *Frankenstein*.
1827—Karl von Baer becomes the first to see a mammal's egg.
1837—Queen Victoria takes the throne.
1830s–1860s—Cell theory emerges.
1861–1865—The American Civil War lasts four long years.
1875—Oscar Hertwig witnesses the union of sperm and egg.

ENGLAND IN
THE EARLY 1630s

I N CENTURIES TO COME THESE FIELDS AND WOODLANDS WILL shrink to tiny patches of green in a vast city. Londoners and tourists will feed ducks and swans here, and pose for giggling pictures. But today there are no crowds, no sightseers, no drifting sounds from the world outside. We are in an English royal park, the property of King Charles I. The king and his physician, William Harvey, are hunting deer. It is rutting season.

Neither Harvey nor the king has heard of a "locked room mystery," where a body is found in impossible circumstances. Perhaps a dead man is discovered in a study locked from the inside, with a knife plunged into his back. Neither man has imagined such a thing. They are about to.

Harvey is a small man with raven-black hair and dark, darting eyes. Ambitious, impatient, and, as a friend put it, notoriously "hotheaded," he radiates intensity. He is destined to soar into the medical pantheon for proving that the heart is a pump that sends the blood circulating around the body through an intricate network of arteries and veins.

Charles is slender, handsome, solemn, utterly convinced that God has set him above other mortals and that "the king can do no wrong." He is destined to die at the hands of the English people, his head chopped off by a masked executioner and then held aloft by the hair while the crowd whoops in glee and gasps in shock at what it has done.

HARVEY HAD PUBLISHED HIS ACCOUNT OF THE HEART IN 1628, A few years before the hunting excursion. The world denounced him. "It was believed by the vulgar that he was crack-brained," Harvey complained, "and all the physicians were against his opinion." For a man as combative as Harvey, that disdain served more as a spur than a rebuke. Harvey remained all his life a staunch "seeing is believing" man. Let others prattle on.

For uncounted ages, the heart had been the seat of the soul and the home of the emotions and insights that set humankind above other creatures. (When we talk today about a "kind heart" or a "cold heart" or speak of learning a poem "by heart," the idioms are fossils of bygone beliefs.) What the sun was to the sky or the lion to the jungle, the heart was to the body. Now Harvey had demonstrated that this noble organ was in truth a wet and slimy machine.

The world would come around to Harvey's view, though not for another two decades. In the end, the admiration would be universal. One dazzled follower would celebrate Harvey in verse: "Thy *Observing* Eye first found the Art / Of all the *Wheels* and *Clock-work* of the *Heart.*"

At the time of his hunting venture with the king, that fame still lies ahead. Harvey is embattled, not acclaimed. But he knows what he has accomplished, even if the medical world has yet to catch on. With the riddle of the heart solved, Harvey has turned his attention to the greatest mystery of all. Since humankind's earliest days, men and women have wondered how new life comes into the world. *How does sex lead to babies?* Harvey intends to find out. He will learn, precisely, how mating creates life.

He will start by studying deer, as a matter of practicality, though humans are the real prize. The king is an avid horseman and hunter who, as Harvey notes happily, "is wont for Recreation and Health sake to hunt almost every week." Harvey has managed to enlist his king as his ally.

The king's huntsmen bring down a doe. Harvey, the most renowned anatomist of the age (and one of the last great anatomists to rely on what he can see with his naked eye), pushes close. Now he will show the king, who "much delighted in this kind of curiosity," the secrets of conception and pregnancy. Together, they will gaze on a deer embryo in its earliest days. They are about to see—what has never been revealed to anyone before—a small, round, glistening globule like an egg without a shell.

Harvey thrusts his knife into the doe's belly and cuts her open. Steam from the hot body rises into the chilly air. Harvey peers inside the animal's womb, first avidly and then perplexedly. The king looks over his physician's shoulder. They see . . . nothing!

No sign of semen, no embryo, nothing whatever to distinguish this doe from any other, though Harvey and all the king's huntsmen have no doubt that she is pregnant. Harvey calls the king in closer and points out that there is "no seed at all residing in their Uterus."

In the coming days, Harvey repeats the procedure again and again, always with the same result. Despite the most careful search, he never sees any semen in these newly mated does; he never sees any odd bits that might represent the female's contribution to conception; he never sees any changes in the deer's ovaries; he never sees any hint of an embryo.

Could it be that Harvey, the king, and the huntsmen have all somehow deluded themselves? Perhaps they have been carefully studying deer who have not mated after all.

Harvey devises a test. This time he will wait for the end of breeding season, when there can be no doubt that he is dealing with pregnant females. He will take a group of females, pick a few at random to dissect, and leave the others alone. (This is another breakthrough: Harvey is one of the first experimenters—by some accounts, *the* first—to use a "control" group.)

Why go to so much bother? Because whatever Harvey sees in the bodies of the cut-open deer, chosen randomly, he would presumably also see if it were possible to peek inside the bodies of the living deer.

With the king's permission, Harvey takes a dozen females and pens them up, so he can keep track of them. He chooses several at random to sacrifice and dissects them. As usual, he finds nothing whatever. Now he waits and watches the remaining deer. At the customary time, they deliver fawns.

None of this makes sense. There can be no doubt that males produce semen. Everyone knows what it looks like. It is a real, physical, commonplace substance. Everyone knows it is essential for pregnancy. How could it be that, when the male impregnates the female, the semen vanishes? Where is the semen? Where is the embryo?

Stranger still, it is fact-minded, dogged William Harvey, the least flighty of men, who has delivered this bizarre news. With his revolutionary picture of the heart, he had showed in the most irrefutable way that he could explain the body's workings in naturalistic, down-to-earth terms. And now here is Harvey himself proclaiming the death of common sense!

The royal gamekeepers, angry and disbelieving, weigh in. "They peremptorily affirm that I was first mistaken myself," Harvey growls, "and so had drawn the King into my error, and that it could not possibly be."

The deer were pregnant. There was no physical evidence whatsoever. It could not possibly be. But it was.

{PART ONE}

PEERING INTO THE BODY

*"It is quite a three pipe problem, and I beg that
you won't speak to me for fifty minutes."*

—ARTHUR CONAN DOYLE, "The Red-Headed League"

ONE

ONWARD TO GLORY

B Y THE LATE 1600S, THE ERA WHEN THE SCIENTIFIC WORLD began to take on its modern shape, explorers had circled the globe and mapped the heavens. They had calculated the weight of the Earth, traced the paths of comets that cut the sky only once in a lifetime, and divined the secret of the Milky Way. They had uncovered the mathematics at the heart of music and discovered the laws of perspective, so that an artist armed only with a paintbrush could pin real ity to his canvas. But for thousands of years, long after Columbus and Magellan and Galileo, the deepest scientific riddle of all lay unsolved.

Where do babies come from? Such geniuses and creators of the modern era as Leonardo da Vinci and Isaac Newton did not know. They knew, that is, that men and women have sex and as a result, sometimes, babies, but they did not know how those babies are created. They did not know that women produce eggs, and when they finally discovered sperm cells, they did not know that those wriggly tadpoles had anything to do with babies and pregnancy. (The leading theory was that they were parasites, perhaps related to the newly discovered mini-creatures that swam in drops of pond water. This was Newton's view.)

Not until astonishingly recent times—in 1875, in a seaside laboratory in Naples, Italy—was the mystery of where babies come from finally solved.

Until then everything to do with conception and development was wrapped in darkness. For centuries, scientists struggled to find out if the woman merely provides a fertile field for the man's seed, or if she produces some kind of seed of her own. They did not know how twins come to be. (Too much semen? Two bouts of sex in quick succession? Sex with two different men?) They did not know if conception is more likely on the night of a full moon or a new moon or if timing makes any difference at all. They did not know, though they assumed, that a baby has only one father, as it has only one mother. They did not know why babies resemble their parents, and sometimes one parent more than the other.

Where do we come from? How does life begin? These were the most urgent of all scientific questions. The world is festooned with mystery and miracle. But not everyone has wondered why the stars shine or why the Earth spins. Every person who has ever lived has asked where babies come from. For millennia, the deepest of thinkers (and every ordinary person) had pondered this cosmic riddle.

No one had a clue.

PART OF THE REASON FOR THE PERPLEXITY WAS STRAIGHTFORward. We tend to forget how astonishing the story of life truly is. We've heard the explanation so often that we take it to be common sense. *Every fourth grader knows where babies come from.* Both parents contribute equally, we learn early on, the mother providing an egg and the father sperm. Each month one of a woman's ovaries releases an egg, which travels through a Fallopian tube toward her uterus. If a couple has sex at the right time, some of the millions of sperm cells in a man's semen make their way from her vagina and cervix toward that egg. One of those sperm cells may fuse with the egg. In time that newly merged cell divides into two joined cells, and then into four and eight and so on. After nine months, a new human being bursts, howling, into the world.

The truth is so far-fetched that it is a wonder that anyone believes it.

In textbook accounts of science, far-seeing researchers systematically gather facts and pile them in sturdy and imposing towers. The story of sex and babies was nothing like that steady advance toward a goal. The scientists who finally solved the case ventured off course for decades at a time. They raced at top speed down long, dark alleys chasing suspects who turned out to have airtight alibis. They concocted elaborate scenarios that collapsed in fantasy. They wandered in a daze, stymied by observations they could not fit into any pattern. They found some clues by deep and careful investigation and others by tripping over them as they raced in the wrong direction in the dark.

Progress came in fits and lurches, but that is the way with all true mysteries. Only in old-school television does insight arrive on cue, just in time for the closing credits. The problem was not that the scientists were incompetent—they were human and fallible, but many were dazzlingly intelligent, and nearly all were diligent—but that the truth was so well concealed.

To crack the case, scientists would need new tools, notably the microscope, and new ideas, notably the insight that the body is made of cells, trillions upon trillions of them, which all arise from a single progenitor. More than tools and ideas, they would need whole new ways of thinking. Suppose, for a moment, that some early savant had somehow leapt to the true conclusion that a living organism begins as a single cell. What then? Immediately scientists would have found their path blocked by a Sphinx posing a bewildering follow-up riddle—how does that single tiny cell "know" how to transform itself into a gurgling, six-pound baby?

Tackling that question would have required these early scientists to understand that a living organism could *assemble itself!* Through most of the history of the world, this was unthinkable, as outlandish as the idea that a cathedral could build itself. Today it is a concept hammered into every student who takes high school biology.

The path to that insight was tangled and difficult. It required, among other things, drawing analogies from machines like player pianos and, later, computers, that carried out complicated actions by following instructions written in code. In the twentieth century, such mechanical devices would lead to the discovery that life itself followed instructions written in a genetic code. But in the 1600s and 1700s no such machines existed; no one could look at the ghostly motions of a player piano's keys, governed by a musical roll, and shout, "Eureka!"

Instead, scientists looked around them in every direction for clues to the riddles of conception and development. *How could it possibly work?* Baffled but determined, they ventured down the most unlikely paths. They studied insects with obsessive care, for example, in the hope that those startling transformations—*a wriggling caterpillar inside its cocoon emerges as a butterfly with gossamer wings!*—would throw light on the changes in infants and babies. They studied fish and frogs and dogs and deer to see what they shared in the way of anatomy and mating behavior. They tackled the narrowest of questions—*How do snails, which have both male and female genitals, sort out who will do what to whom?*—and the grandest of themes—*Do living organisms possess a "vital force" that sparks them to life?*

Often a quest that started in one direction ended far afield, in a landing spot no one had anticipated. The search for the vital force, for instance, led to strange and dangerous experiments with electricity and lightning, and even an encounter with Dr. Frankenstein and his monster.

In hindsight, that zigzag progress seems inevitable, for two seemingly different questions were deeply entangled. The question, *Where do babies come from?* proved impossible to separate from *What is life?* Thus, straightforward inquiries about sperm and eggs and anatomy opened up into profound riddles about the nature of living organisms. Scientists who merely wanted to know about the body's nooks and crannies found themselves wondering what the world was made of.

How could it be that the same, ordinary stuff that forms lumps of mud and murky puddles, when rearranged in the right way, can burst into crying, crawling life?

This was, moreover, an era when science and religion wrapped around one another, so that every statement about life implied a judgment about God the Creator. With one peek through a microscope, a scientific observation could provoke a religious battle.

The search for the solution to the sex and conception mystery would consume careers and span centuries. My focus is on the heart of the tale, from about 1650 to nearly 1900. During that time, scientists in half a dozen countries—in Holland, France, England, Germany, Italy, even the newborn United States—would pass the torch, or burn their hands and drop it. It is surprising, in hindsight, that it did not take them *longer*.

T HEY SET OUT, NOT WITH HESITANCY OR TREPIDATION, BUT WITH eagerness and swagger, for their peers had just won a colossal victory. In a torrent of discoveries all through the 1600s, physicists and astronomers had demonstrated that the universe was a clockwork, the stars and planets its cogs and wheels. Spirits and demons had been banished. Nature obeyed laws and equations, not caprice. Comets were not heavenly messengers. God was a mathematician.

The seventeenth century was a tumultuous age, bubbling over with intellectual ambition. The roll call of genius went on and on. Shakespeare, Galileo, Rembrandt, Descartes, Newton were not busts on a library shelf but flesh-and-blood titans who wielded almost unfathomable powers. The secrets of the human heart and the natural world seemed destined to give way before them.

The scientists in that starry roster shared not merely an ambition but a way of thought. From Galileo and Newton on, they believed with an unshakable faith that their mission was to study God's creation and proclaim the glory of his craftsmanship. This they had begun to do, and when they rattled off the list of their triumphs they were making a twofold claim. First, they had done great things. They had unraveled the rainbow and harnessed the tides. Second, and just as important, they had cracked a cosmic code. They had found out *how* to think of the world: it was an encrypted message written by God.

The world was not just beautiful, like a painting that moved onlookers to delight; it was a beautifully constructed riddle that contained secrets within secrets. Humankind's highest task, these early scientists fervently believed, was to solve that riddle, the better to honor the Creator. (And why had God made matters so difficult? If he wanted trembling homage, why not arrange the stars to spell out BEHOLD in letters of fire? For the devout scientists of the 1600s, the answer was plain. God had given humankind its defining gift, a far-ranging intellect, and he meant for us to use it.)

The book of nature was written in the language of mathematics, Newton and his contemporaries insisted, and only the mathematically literate could read it. This was an austere view. Romantic poets would later howl that the scientists' "cold philosophy" had drained the world of its sap and energy, as if a lush Gauguin landscape had been set aside in favor of a diagram in a geometry textbook.* But despite their somber mathematical talk, the scientists of the seventeenth century tackled the world with gusto and high hopes.

They talked of God the mathematician, but they seemed to picture him in less imposing terms. Francis Bacon, one of the pioneers of the Scientific Revolution, took this to an extreme. His God seemed less the Almighty on a celestial throne than an indulgent parent who had devised an Easter egg hunt for a pack of toddlers. God "took delight to hide his works," wrote Bacon, "to the end to have them found out."

Bolstered by their successes in understanding the physical world, scientists made bold plans to continue their advance, this time moving to living creatures. If the proud heart could be understood as a pump, the rest of the body seemed sure to follow. Muscles and bones and

---

*Perhaps it is not surprising that God as the mathematicians pictured him bore a striking resemblance to the mathematicians themselves. "If triangles had a god," Montesquieu would write a few decades later, "he would have three sides."

arteries would play the roles of levers, pulleys, and pipes; food would serve for fuel; filters and springs, plumbs and bellows, would all have their natural counterparts.

In time, the body would yield its secrets like a watch opened up on a craftsman's bench. Humans and animals would show themselves as more wondrous versions of the ingenious mechanical figures that adorned the clock towers of Europe's cathedrals and emerged, on the hour, to march a few proud steps and raise trumpets to their metal lips.

The inner workings of life, the deepest mysteries of how animals move and breathe and reproduce, would be unveiled. God, who had placed the stars in the sky and carved towering mountains, had surely lavished even greater care on the details of a bird's wing and a whale's fin and, especially, on the human beings he had fashioned in his own image.

Scientists pictured the living world as a straightforward target. The inanimate world, with its hurtling planets and remote, untouchable stars, seemed more forbidding, and, indeed, Newton and his peers had needed to invent a mathematical language to describe it. Even today, we think of physics, with its black holes and hidden dimensions and parallel universes, as the realm of the incomprehensible. Biology, in contrast, deals with everyday, literally down-to-earth objects like dogs and plants and snails. Scientists had conquered physics even so. Biology would fall, too. They would make quick work of it.

They didn't. The bold men of science raced off to take on the mystery of life and promptly face-planted. Life was messy, it turned out almost at once, fragile and unpredictable and maddeningly coy about its secrets. A cannonball in flight plainly has no desire to go this way rather than that. Why shouldn't a few simple laws suffice to describe its motion? But consider the gulf between a chunk of metal and a sunflower craning toward the light, or a dog tugging on its leash, or a human being. On the one side, mere matter. On the other, breathing, pulsing life. How *could* the same sort of laws account for both?

Planets were easy; plants were hard. And sex and reproduction proved hardest of all. For nearly three hundred years, from about 1600 until nearly 1900, the question of where babies came from defeated one thinker after another. The vanquished lay in heaps, like soldiers on a battlefield. Late in the 1700s one scientist took the trouble to compile a list of failed theories. Scanning the centuries, he tallied "262 groundless hypotheses." ("And nothing is more certain," one eminent biologist remarked at once, "than that his own system is the 263rd.")

TWO

HIDDEN IN DEEP NIGHT

W E KNOW HOW THE SEX-AND-BABIES STORY TURNED OUT. THAT can make us smug—*how foolish of our forebears to have lived so long ago*—but it shouldn't. To understand the plight of our predecessors, think of modern-day scientists as they wrestle with their own grand mysteries. The hardest problem in science today—so difficult that scientists refer to it simply as "the Hard Problem," as if there were no other—is to explain perhaps the simplest fact in the world. Why is it that it feels like something to be us? Why is it that a robot, even one that can find its way around the room and beep and blink and play chess, is simply a lifeless collection of parts, whereas we humans swim in a sea of smells and sights and memories? What, in short, is consciousness?

If every object in the world is just a collection of atoms lumped together, how is it that some lumps just sit there but the three-pound lump that is our brain conjures up a world? How can mere *stuff* do that? In the words of the poet and essayist Diane Ackerman, "How do you begin with hydrogen and end up with prom dresses, jealousy, chamber music?"

Today the deepest thinkers in the world can only stammer in response. One day, the answer may be so obvious that nobody will understand how there could ever have been any confusion. In the future, nine-year-olds may read books called *Where Ideas Come From*.

In centuries past, the riddle of life seemed every bit as intractable as the riddle of consciousness does today. We understand perfectly well that brain gives rise to mind; the problem is that we cannot sort out just what that means. The scientists who made the modern world understood that certain bits of matter were alive and others weren't; the problem was that they couldn't sort out how that could be.

We cannot grasp how the brain, a lump of meat locked inside a dark, bone-framed cave, can create a light-soaked world. *They* could not grasp how a few inert odds and ends could take the shape of a leaping tiger with daggers for teeth. To try to make headway by focusing on what seemed a narrower question, *Where do babies come from?*, turned out in fact to make matters even harder. Tackling the riddle of life was challenge enough; explaining *new* life was harder still. Within a woman's womb a microscopic bit of tissue grows into a baby. But how did it get there? It couldn't appear out of nothing. Where did it come from in the first place, and then how did it grow?

M OST CONFUSING OF ALL, TO THE MALE SCIENTISTS CONTEM- plating these riddles, was sorting out the woman's role in this story. Plainly she carried the baby and delivered it, but what did she contribute to the *making* of it? The male produced semen. What did the woman do? In the twentieth century Freud would famously ask, "What does woman want?" In the seventeenth century the question was, "What are women for?"

The simplest answer, a favorite of male thinkers since ancient times, was that woman was the field where a man planted his seed. That view was always presented as if it were the merest common sense. Aeschylus had spelled it out four centuries before Christ.

The woman you call the mother of the child
Is not the parent, just a nurse to the seed,

the new-sown seed that grow and swells inside her.
The *man* is the source of life—the one who mounts.
She, like a stranger for a stranger, keeps
the shoot alive unless god hurts the roots.

In England twenty centuries later, many still clung to the same view. The king's royal physician put it briskly, in a book on anatomy published in 1618: "The woman hath a womb ordained by nature as a field or seed-plot to receive and cherish the seed." But confront those learned authorities with everyday observations, and their certainty vanished. *If the mother is merely a field where the infant grows, why do children so often look like their mother?* That hit close to home. Maybe women *did* shape their babies in some way? On the one hand, that seemed unlikely, given women's second-class status. On the other, it seemed indisputable, for how else to explain family resemblances. What was going on?

FROM A DISTANCE, SEX LOOKS SIMPLE ENOUGH, A BIT OF HUFF-ing and puffing and some rudimentary choreography. But the key scenes in the drama—conception and then development over the course of nine months—take place deep inside the body, hidden from view. Nature, lamented William Harvey, concealed these biological secrets "in obscurity and deep night."

And even if you could see, you could not expect much from a search where you were not certain what you were looking for. No one knew, for instance, if women had eggs. Scientists split into feuding camps—one insisted that women, like birds, produced eggs; their rivals shouted that women, like men, produced a sort of semen.

Opinions abounded, but facts were rare and elusive. Egg and sperm, we now know, were not merely hidden but tiny. The human egg, though it is the largest cell in the body, is only the size of the period at the end of this sentence. Sperm cells are the *smallest* in the body, far too little to see with the naked eye. (An egg outweighs the sperm cell that fertilizes it by a million to one, the difference between a Thanksgiving turkey and a housefly.) The mystery of human development was never going to be as simple to investigate as the path of a falling rock.

Merely sorting out the basic facts of human anatomy was hard, grim work. To peer inside the living body was next to impossible.* Dead bodies were the only alternative but a poor fallback when your goal was understanding life. Dissections were carried out in the cold (to keep the corpse from decomposing) while the anatomist poked his knife into dark, wet crevices. Fascination and horror twined around one another. "You might be stopped by your disgust," Leonardo da Vinci wrote, no matter how strong your curiosity, "and if that did not hinder you, then perhaps by the fear of spending the night hours in the company of those dead bodies, quartered and flayed and terrifying to behold."

High as the practical hurdles were, others stood higher still. All questions that had to do with birth and babies were charged and not simply because they had to do with that most fraught of all subjects, sex. In a God-drenched age like the 1600s, a venture into science was a dive into religion. Those were roiling waters. Everyone believed that God had fashioned the world and all its inhabitants, as the Bible detailed. God alone had the power to create life. How could one speak of the exalted work of creation and, in the same breath, of ordinary men and women clutching one another and gasping in the dark?

The whole scientific enterprise, with its talk of humans as sophisticated machines, was dangerous in two different ways at once. First, it threatened to push God to one side. That was blasphemy, and no more hideous charge could have been imagined. Second, and almost as bad, the scientific way of thinking seemed to drain purpose and meaning from the world. An assemblage of lifeless parts was not responsible for its actions. If humans were machines, a husband might strangle

---

*But not *altogether* impossible. In 1640 William Harvey met a young man who had fallen from a horse onto a sharp rock; he had been left with a hole in his chest, which he covered with a metal plate. He had fully recovered, and his injury made him a celebrity. (Once, in Rome, a crowd eager for a peek inside a living body had filled an opera house to see him.) Both Harvey and then King Charles I eagerly reached inside the man's chest and placed their fingers on a beating human heart.

his wife and bear no more blame than a runaway carriage that plowed into a crowd.

In a different age, those contrasting approaches to the world would have pitted scientists against religious believers. That conflict would have been clear-cut. But in this case the scientists and the believers were the same people. The battle lines ran not between opposing factions but through the minds of individual men. Brilliant, ambitious, confused, conflicted, these reluctant revolutionaries sought desperately to find a way to fit their new discoveries with their old beliefs.

When it came to sex, they found themselves more confused than ever. They could not imagine, first of all, why God had devised so bizarre a system for preserving the human species. What could be less dignified? "Who would have solicited and embraced such a filthy thing as sexual intercourse?" demanded France's royal physician, in 1600. "With what countenance would man, that divine animal full of reason and wisdom, have handled the obscene parts of women, befouled with such great quantities of muck and accordingly relegated to the lowest part of the body, the body's bilge as it were?"

And, the learned physician went on, women didn't fare so well out of this sex business, either. "What woman would have rushed into a man's embrace unless her genital parts had been endowed with an itch for pleasure past belief? The nine months of gestation are laborious; the delivery of the fetus is beset with dreadfully excruciating pains and often fatal; the rearing of the delivered fetus is full of anxiety."*

In the end, though, it was God who had ordained this odd system, just as he had created the sky and the seas. No doubt he had his reasons.

THE GREAT INVESTIGATION STARTED WITH A FEW FACTS THAT EVeryone could agree on, but these were isolated landmarks in a vast

---

*Scientists today still marvel (or cringe) when they consider how lightly men get off in the actual labor of baby making. The evolutionary biologist Robert Trivers contrasts "a sperm cell weighing one-trillionth of a gram" with "a nine-month pregnancy producing a seven-and-a-half-pound baby."

and empty landscape. Nearly every culture had figured out early on that it takes two to tango. Of the two partners, males presented fewer mysteries than females because they kept their working parts on the outside, on conspicuous display, in the mode of the Pompidou Center.

The role of the penis seemed plain enough, though just how it managed its furlings and unfurlings was not resolved. Biologists and physicians all agreed that the testicles had something to do with sex and babies, too, but it was by no means clear what they did. (Aristotle contended that they were merely oddly packaged counterweights, akin to those that women hung from the threads in their looms to keep them from tangling. In the case of males, Aristotle explained, the weights served to keep the seminal vessels unsnarled.)

Semen, as the only impossible-to-miss product of sex, was plainly a crucial part of the puzzle, but it remained utterly mysterious. The standard notion among scientists (virtually all of them male, in these early days) was that semen was a magical, almost divine concoction. Precisely what form that magic took was in dispute: Did semen exert its influence without physical contact, as sunlight nurtured plants, or did it serve as the key ingredient in a divinely ordained recipe, as a kind of baby batter?

Women's anatomy was poorly understood. The structure of the vagina was known, and so was that of the uterus. After that, almost any question would have elicited confusion or dispute. What were the Fallopian tubes for? What about the ovaries, which did not even have a name of their own and were referred to as "female testicles"? What was menstruation? Beyond agreement across cultures and across millennia that it served as proof of women's low status (menstrual blood soured wine, withered grass, and drove bees from their hives, according to Pliny, the acclaimed Roman author of *Natural History*), no one knew.

Heredity was an especially perplexing riddle, even in its most rudimentary aspects. No one could explain why horses gave birth to colts, or why dogs had puppies rather than kittens. Look closer, and the mystery only grew deeper. Embryos from different animals looked much like one another and not like much of anything. How did one tiny clump of tissue know to grow into a kitten and another, which looked nearly identical, into a calf? The facts were so familiar—"like gives rise to like" was an ancient observation—that scarcely anyone had ever pondered them. But once such questions were raised, they met only head-scratching and double talk.

Other questions, just as basic, met equal befuddlement. If babies somehow combined features of their two parents, as experience seemed to demonstrate, how was it that newborns were either boys or girls rather than a combination of the two? And if the two parents each contributed to forming their baby, why weren't babies born as monsters with two heads and four arms and four legs?

THESE QUESTIONS WOULD HAVE PROVED DIFFICULT NO MATTER who took them on. But virtually without exception, the scientists wrestling with these mysteries were men. More than that, they were men who took for granted that women were their physical and mental inferiors. Not all of them would have gone as far as Aristotle, who described females as "mutilated males." But the scientists' goal was understanding how men and women, together, create babies. To start with the assumption that one of the two participants wasn't up to much was to ask for trouble.

Take the vexed matter of eggs. They had always been associated with new life, presumably because everyone had seen tiny birds peck out of their shells. Countless cultures told creation myths about how the first humans had emerged from an egg. Ancient Indians and Chinese and Tibetans and Celts believed that all of heaven and earth, and not merely human beings, had come from a cosmic egg.

In the seventeenth century, scientists found still another reason to look with special favor on eggs and ovals of all sorts. God the mathematician, they declared, had favored the circle above all other shapes, because it was geometrically perfect. (William Harvey's confidence in his picture of the blood circulating through the body rested in part on this faith in God's fondness for circles.)

In the heavens, God had proclaimed his devotion to circles in a kind of cosmic calligraphy. The simplest and most elegant shape, the embodiment of eternity, a curve without beginning or end—no wonder that the greatest of all geometers had made the Earth and the other planets round, and had sent them spinning around the round sun in immense and sweeping ovals.*

So it seemed a good bet that eggs would come into the story of life in a central way. Though no one had ever seen a human egg, many early scientists felt sure they would find one someday. But one strange contradiction bewildered them. Eggs were special; women were not. "What was God trying to achieve through this mixed message?" in the words of the historian Clara Pinto-Correia. "Why would he encase us inside the shape of perfection only to lock that shape within imperfect bodies?"

With so many questions unanswered, all was murk and confusion. "In brief," wrote a French historian of science in a magisterial overview of the early days of biology, "nothing was certain and nothing was uncontroversial, except what was blindingly apparent."

For men of vaunting intellect and ambition, this was maddening. Intent on solving a mystery for the ages, they found themselves in the position of detectives stymied by a killer who mocked their stumbling efforts with taunting notes and brazen challenges. They stood, staring, in front of a wall filled with drawings of suspects and crime-scene photos. Arrows zigged back and forth. Scrawled labels—SUSPECT? BYSTANDER?—marked several images. Here and there an old label had been scratched out and a new, hopeful guess put in its place.

The detectives stepped back and scanned the puzzle pieces yet again, in hope that they had missed a crucial connection. Somewhere in this welter of evidence and guesses they must have overlooked a clue.

---

*Scholars debated why God had defaced the round Earth with mountains, like warts on a perfect face. Until the mid-1600s, mountains, even the Alps, were denounced as "deformities," "boils," and "monstrous" growths. The consensus was that God *had* made the Earth a perfect, utterly smooth sphere. Mountains rose up later, when Adam and Eve were banished from the Garden of Eden. Not only was humankind punished for disobeying God, but so was the Earth itself.

THREE

SWALLOWING STONES
AND DRINKING DEW

A ROUND THE WORLD, LONG BEFORE SCIENTISTS SET OUT TO explain sex and conception, healers and shamans and ordinary people had come up with their own answers to these universal riddles. For a few minutes, let's look not only at Europe in the Age of Science but also at other lands and other eras.

With many of history's great riddles, China or India or another non-Western culture made huge breakthroughs that Europe only caught up with centuries and centuries later. That was the pattern in astronomy and mathematics and geography. It was *not* the story of sex and conception.

When it came to babies, the focus nearly everywhere was on such practical matters as trying to devise potions that led to pregnancy or prevented it.* More theoretical questions—*Where does the baby come*

---

* In the ancient world, contraception was nearly always regarded as solely the woman's responsibility. To prevent pregnancy, women gulped down potions made from elaborate recipes or inserted pastes and salves in their vaginas. Men had it easier. Surprisingly, condoms did not come into use in ancient times. The definitive history of contraception states flatly that "there is little if any evidence for usage of a condom or sheath during antiquity to prevent conception." By the 1700s, men did sometimes wear condoms (in his memoirs, Casanova refers to a condom as an "English riding coat"). But condoms remained rare until scientists learned to vulcanize rubber, in 1844.

from, exactly?—had no such urgency. Just as people around the world built homes and lofty palaces long before anyone had devised a theory of gravity, so people courted one another and made love without feeling any need for a full-fledged theory of conception.

The most pressing question was why some sexual encounters led to pregnancy and others did not. At some point or other, nearly everything was pinpointed as *the* vital factor behind conception: sunlight, moonlight, rainbows, thunder, lightning, rain, a cobra's hiss, the aroma of cooked dragon's heart.

Countless stories revolved around eating particular foods or, sometimes, choking down things that weren't food at all. In both China and Italy, eating flowers led to pregnancy. (Italians favored roses; a red rose brought a boy, a white rose a girl.) In China swallowing a stone or a pearl or drinking dew brought babies; in Ireland, drinking saint's tears; in India, accidentally ingesting crane's dung.

Contraceptive techniques showed similar inventiveness. In ancient Rome, one preparation—this one was wrapped up in a bit of deer hide and tied to the body, not swallowed—was made from worms that lived in the head of a particular species of hairy spider. (The advice came from Pliny, whose wisdom was much esteemed. Even so, it does not sound like a recipe to entice the timid, especially since it appeared in a chapter of Pliny's *Natural History* that dealt mainly with spider bites.)*

In Egypt, one recipe for contraceptives called for "feces of crocodile, smashed up with fermented dough." The recipe, from 1850 BCE, explains that the concoction should be shaped into a pellet and used as a vaginal suppository. Arabic medical texts never mentioned crocodiles but did often recommend contraceptive suppositories made from elephant dung.

For aphrodisiacs, too, no procedure was too difficult. One Egyptian papyrus now in the British Museum listed the ingredients of a potion sure to win a woman's love: "Take dandruff from the scalp of a dead man, who was murdered," the instructions begin. "Add the blood of a

---

*We met Pliny before, warning that menstrual blood had the power to turn wine sour and drive bees from their hives.

tick from a black dog, a drop of blood from the ring finger of your left hand, and your semen."

MILLENNIA BEFORE THE EGYPTIANS, OUR EARLIEST FOREBEARS would have made their own guesses at the sex and babies mystery. Expert observers of the natural world, they would have known the subtlest properties of their surroundings: which plants were good to eat, which reeds made the best baskets, which vines could be spliced into rope.

Many of nature's features would not have been subtle at all. The sun traveled dependably across the sky; the moon changed from fingernail sliver to gleaming disc; gray skies poured down rain; the night echoed with howls and roars. Among these boldfaced scenes, one stood out. Some young women watched their bellies swell and then, months later, pushed a tiny, flailing newcomer into the world. This was bizarre and paradoxical: How could an event be both common and miraculous?

Sorting out how such a thing could be and what it had to do with sex (if there was any connection at all) presumably took a long while. Certainly there were clues. We know that no culture can have failed to discover intercourse. The temptation is to picture happy, dazed couples bursting out of the bushes, exuberantly high-fiving, as if they had managed the prehistoric equivalent of putting together a table from IKEA instructions. Someday, it seems, we will unearth an Inventors Hall of Fame. There homage will be paid to the benefactors of the human race who discovered sex and fire and storytelling.

But that cannot be right. Humans would always have known about sex, as part of their genetic legacy. Just as babies take their first steps without advice from their parents ("now shift your weight to your forward foot"), humans would have taken to lovemaking as naturally as to laughing or talking. They would have started tallying up clues to the baby riddle early on. Women who did not have sex with men did not have babies, first of all. And the place where the penis entered the mother's body, months ago, was the very place where babies came out, today! That was circumstantial evidence, not proof, but it would have made a person think.

So would the mating behavior of animals. But the connection with humans would not have been self-evident, since many animals mate only at a specific season of the year, with pregnancy and then the birth of the new generation following in lockstep. With humans, the pattern of cause-and-effect would have been harder to spot since sex (and therefore birth) can occur at any time whatever.

Infertility muddied the picture. Even the great many cultures that took for granted that sex played a key role in baby making found themselves thinking that sex could not be the whole story. Perhaps the childless had offended the gods, or eaten the wrong foods, or gone to bed at the wrong time or in the wrong place or in the wrong frame of mind.

N O COUNTRY HAD A MONOPOLY ON MISGUIDED SEXUAL THEORIES, but the path of misinformation is easiest to trace in Europe. Books offering sexual advice were among the earliest printed offerings, and they flew off the shelves. Perhaps the most popular of all was *Aristotle's Masterpiece*, first published in 1684.

Just as the Holy Roman Empire was, according to Voltaire, neither holy nor Roman nor an empire, so *Aristotle's Masterpiece* was neither by Aristotle nor a masterpiece. But it offered advice to those who were "fond of nocturnal Embraces," and readers grabbed it up. New editions appeared as late as the 1930s. (In *Ulysses*, James Joyce describes Leopold Bloom flipping through its pages.)

The *Masterpiece* tended to the vague (couples should "survey the lovely beauties of each other") and the mistaken (the man "ought to take care not to withdraw too precipitately from the field of love lest he should, by so doing, let the cold into the womb, which might be of dangerous consequence"). When the text was not coy, it was lurid. Woodcuts showed such "monsters" as a boy covered with fur and

FIGURE 3.1. This hugely popular sex manual, *Aristotle's Masterpiece*, appeared in hundreds of editions over the course of 150 years. The cover shows a red-clad scholar, supposedly Aristotle, and a half-naked woman meant to entice the reader.

twins joined at the shoulder, and the book warned that these were the penalties paid "by the undue coition of a man and his wife, when her monthly flowings are upon her; which being a thing against nature, no wonder that it should produce an unnatural issue."

Certainly until the 1700s, that tone of authoritative ignorance was the hallmark of all theorizing on sex and pregnancy. In England and continental Europe, for instance, the question of just what was natural and what was not provoked endless debate. Folk wisdom taught that woman-on-top was a sure way to avoid pregnancy. (A popular poem chastised "Subtle Lechers! Knowing that / They cannot so be got with Brat.") The medical community could not agree on whether this form of lovemaking was a valid anti-pregnancy strategy, or even if it was ethical. One well-regarded French physician warned that any children conceived in this way would likely be "Dwarfs, Cripples, Hunch-backed, Squint eyed, and stupid Blockheads, and by their Imperfections would fully evidence the irregular Life of their Parents."

Theologians weighed in, generally on the side of those who were outraged at the depravity all around them. Here truly was an example of women who did not know their place. "When the woman is on top, she acts [rather than 'accepts']," one scholar scolded. "Who cannot see how horrified nature is by this aberration?" Another religious writer explained that "the cause of the Flood was that the women, overcome with madness, had misused the men, the latter being underneath and the women on top."

E LSEWHERE IN THE WORLD, AND INTO MODERN TIMES, A HAND-ful of cultures seem never to have bought into the idea that babies come from sex, at all. The most famous holdouts were the people of the Trobriand Islands (which are today part of Papua New Guinea). In the early decades of the twentieth century, according to one of the renowned figures in the history of anthropology, the Trobrianders remained "entirely ignorant" of the connection.

The Trobriand explanation of where babies come from, Bronislaw Malinowski learned, was complex. When men and women die, their soul or spirit, the *baloma*, travels to an island called Tuma, about ten miles northwest of the Trobrianders' home islands. (This was a real place, inhabited by living people as well as spirits.) On Tuma the spirits settle down with the *balomas* of their relatives and go on with their afterlife. They eat, sleep, age, fall in love. Eventually they grow old. Then the *baloma* goes down to the beach, wriggles out of its ancient skin, and transforms into a tiny embryo. These embryos, which the Trobrianders call "spirit children," are the key. At some point a young woman bathing in the sea will feel something touch her. "A fish has bitten me," she may exclaim, but in truth she is now carrying a spirit child.

In keeping with this theory, the Trobrianders had no notion of paternity. (Their word for "father," according to Malinowski, translated as "husband of my mother.") Incredulous and perplexed, Malinowski asked the islanders endless questions. If he named an unmarried woman and asked who was the father of her baby, he met only puzzled stares and the repeated message, "It is a *baloma* who gave her this child." In cases when a man returned from an absence of a year or two to find his wife pregnant, Malinowski learned, no one reacted with anger or dismay.

He changed tack, launching into a simile about planting a seed in the ground and watching it grow. "They were curious, indeed, and asked whether this was 'the white man's manner of doing it.'" It was emphatically not their way. Nor did they find anything compelling in Malinowski's questions about semen. Yes, of course, there was such a thing, but it served the purposes of pleasure and lubrication. Both sexes produced fluid, and the Trobrianders used the same word for both semens.

Finally, though, Malinowski thought he had made a breakthrough. If a girl had not had sex, the Trobrianders explained, she would not have a baby. Happy at last, Malinowski rattled off more questions. His good cheer vanished. It turned out that sex *did* have something to do with babies, but only in a limited, mechanical sense. The reason virgins did not conceive was that their vaginas had not been "opened up." The role of sex was to perform this widening, so that, at some future time when a woman bathed in the sea, a spirit child might enter her. The more sex, the more widening. Now that they had finally managed to convey this information to him, the Trobrianders went on, Malinowski could understand several important aspects of pregnancy. *Now* did he see why it was that virgins never got pregnant, and why women who rarely had sex rarely did, and why women who often had sex were nearly sure to find themselves pregnant?

Malinowski shifted ground. What about animals? Trobrianders raised pigs. Did pigs have their own *balomas* who brought little piglets? The islanders scoffed. The white man with the foolish questions truly was a simpleton. What did animals have to do with anything? To ask questions about how they reproduce would be to speculate on the afterlife of dogs and pigs, and that was a topic that no sensible person would bother with.

The Trobrianders had questions of their own for Malinowski. Sex was an incredibly common feature of life. How did Malinowski explain "that the very act which a woman performs almost as often as eating or drinking, will, once, twice, or three times in her life, cause her to become pregnant?"*

For a century, ever since Malinowski met the Trobrianders, anthropologists have fought over accounts like his. The jury is still out. Do these stories truly reflect what people in New Guinea and Australia and elsewhere believe? Could the locals have been pulling the legs of the earnest anthropologists with their notebooks and endless questions? Or were they professing "official" beliefs rather than their own views, as a halfhearted churchgoer in the west might declare that Jesus was born to a virgin? Or perhaps stories about young girls impregnated by fish bites offered a way for cuckolded husbands and unfaithful wives to save face?

IT MIGHT SEEM PLAUSIBLE THAT SOME CULTURES HAVE NOT caught onto the father's role in baby making. When the baby arrives, after all, the mother is center stage, and the father may be long gone. More surprisingly, cultures separated by thousands of miles and thousands of years have independently reached the conclusion that *mothers* have no biological connection with their own children. Some of these cultures exist today.

This takes some fancy footwork. The customary strategy is to paint the mother as a glory hound. She delivered the baby, true, but she had nothing to do with *creating* it, which was the real achievement. That was the work of spirits, or gods, or the father. The mother was an incubator with delusions of grandeur.

---

*The Trobrianders did not practice birth control, of course, and anthropologists have not resolved the mystery of why pregnancies were not more common. Perhaps it is worth noting that wild yams, which were a staple of the Trobriand diet, contain a hormone that served as the basis for the first birth control pill. (But no modern studies have demonstrated that yams have any contraceptive effects.)

FIGURE 3.2. The Sun God creating the universe. The drawing, from about 1000 BCE, is a detail from a papyrus now in the British Museum.

In ancient Egypt, the creation of new life—indeed, the creation of the entire universe—was emphatically the province of males. Females played a subsidiary role or (in the case of the gods) no role at all. Creation myths told of male gods who, as one historian writes, "gave birth to their spouses, their children, other humans, animals, cities, sanctuaries, shrines, perpetual offerings, earth, and the planets themselves."

One papyrus manuscript records the boasts of the Sun God, who first created himself out of nothing—we are not told how—and then took matters into his own capable hands, masturbating the universe into existence. "I created on my own every being . . . my fist became my spouse. I copulated with my hand."

A second papyrus depicts a variant of the same legend, in which the Sun God again coaxes the universe into being, albeit in a slightly different manner (see Figure 3.2).

Again and again, cultures that had nothing in common came up with nearly identical answers to the riddles of sex and babies. Few cultures can have shared as little as the ancient Greeks and today's African bushmen, for instance, but, as the anthropologist Lorna Marshall writes, "the !Kung believe that in conception the woman's menstrual blood unites with the man's semen to form the embryo." This was precisely Aristotle's view. (He reasoned that menstrual blood must serve some important function, noted that pregnant women do not menstruate, and leapt to the wrong conclusion.)

With yet another riddle—*What happens in the embryo's very earliest days?*—we find remarkably similar accounts in the Book of Job and in

Aristotle. "Didst thou not pour me out like milk and curdle me like cheese, clothe me with skin and flesh and knit me together with bones and sinews?" asked Job. A century or more later, the same cheese making, curdling imagery turned up in Aristotle. When semen and menstrual blood meet, he wrote in *On Generation*, the semen "acts in the same way as rennet acts upon milk."

People as different as the modern-day Basques of Spain and the Bantu of southern Africa, or the ancient Hindus and Hebrews, all hit on virtually the same explanation of how the red and the white parts of the body come to be. "The father provides the white seed, from which are formed bones and nerves, the nails, brain, and the white of the eyes," the Talmud declares. "The mother provides the red-seed, from which are formed the skin and the flesh, the hair, and the black of the eyes." Indian medical writings asserted the same equation— man = white, woman = red—in almost the same words.

Perhaps the most widespread of these shared beliefs was that the man's role in sex was to plant a seed and the woman's role was to nurture it. Seed-and-field imagery dates to ancient times. The Bible is dotted with examples. In Genesis, God commands Abraham to sacrifice his son. When Abraham takes a knife to Isaac's throat, God calls him off and rewards him for his obedience. "I will multiply thy seed as the stars of the heaven, and as the sand which is upon the sea shore; and thy seed shall possess the gate of his enemies."

The historian (and embryologist) Joseph Needham found similar accounts in ancient texts in Egypt and India, and in the Talmud. Needham also cited grim evidence of a different sort that points in the same direction. The belief that women are merely the field for the new generation but do not shape it, he wrote, fits with the widespread practice, in warfare, "of putting captured males to death and retaining the females as concubines. On such a theory, no fear would be entertained of corrupting the race with alien blood in this way."

Everyday English vocabulary contains a less gory hint that the "seed" theory was widespread: the word "semen" comes from the Latin for "seed." The picture of seeds and fields persists to this day. Anthropologists in Turkish villages in recent decades gathered many such accounts. "If you plant wheat, you get wheat," one Turkish woman explained. "If you plant barley, you get barley. It is the seed which determines the kind of plant which will grow, while the field nourishes the plant but does not determine the kind. The man gives the seed, and the woman is like the field."

In modern-day Egypt, as well, poor, urban women still downplayed their own role in conception. "Here in Egypt, we say that the woman is just a container," one woman told the anthropologist Marcia Inhorn. "It is something from God, but she is only a container." The most common view she encountered, Inhorn writes, is that "men bring life and women receive it." (Even so, infertility is always the woman's problem—unless the man cannot have sex at all—because his only task is to ejaculate. He throws; she fumbles.)

Common across many cultures, too, even today, is a belief that it takes many acts of sex to create a baby. "Many of my New Guinea friends feel obliged to have regular sex right up to the end of pregnancy," writes the scientist Jared Diamond, "because they believe that repeated infusions of semen furnish the material to build the fetus's body."

A virtually identical theory is common on the other side of the world, among Indian tribes up and down the South American continent. Though they live thousands of miles from one another and do not interact, many of these far-flung tribes hold to the same little-by-little theory. In one anthropologist's paraphrase, "the fetus is built up gradually, somewhat like a snowball."

In the rain forest, evidently, you *can* be a little bit pregnant. In order for a pregnancy to "take," the fetus must be regularly doused with fresh semen. So demanding is this task that Yanomami men talk about how they have grown thin from their baby-making labors.

Many South American tribes go a step further: not only is the developing baby built up from new batches of semen, but it is best if several different men make a contribution. All those men are considered the child's father. Among the Bari people in Venezuela, for instance, "a good mother will make a point of having sex with several different men, especially when she is pregnant," one historian writes, "so that her child will enjoy the qualities (and paternal care) not merely of the best hunter, but also of the best storyteller, the strongest warrior, and the most considerate lover."

WHEN IT CAME TO SEX, NOT EVERYTHING WAS CONSENSUS. Traditional Jewish and Christian views differed sharply, for instance. Jewish doctrine was far from pro-female (the daily morning prayer of an Orthodox Jewish man includes the words "Blessed are you, Lord, our God, ruler of the universe, who has not created me a woman"), but sex was regarded as something for both partners to enjoy and cherish. Husbands had a duty to provide their wives not just with food and clothing but also with sex. The Talmud even spelled out a schedule: men of means should go to bed with their wives every day, laborers twice a week, camel-drivers once a month, sailors once every three months.

Christian doctrine took a different tack. Sex was suspect. Theologians wrote endlessly on sex and morality, poking their heads into the marital bedroom and occasionally even peeking beneath the sheets. "Shameful kissing and touching" could be condoned, they decreed, so long as the partners hoped to get pregnant and were not indulging for the sake of pleasure. Even within marriage, sex was regarded with suspicion, out of fear that it might divert the participants from spiritual thoughts. "Adulterous is also the man who loves his wife too ardently," the church decreed.*

Religious dogma was important, because science and theology in the West were completely intertwined. Galileo had brought the Inquisition down on his head for suggesting that the earth moved around the sun. His trial showed that there were no exclusively scientific

---

*This view was a long time dying. In 1656, in Boston, a ship's captain named Thomas Kemble returned from a three-year sea voyage and kissed his wife at the doorstep of their home. Kemble was charged with "lewd and unseemly behavior" (made all the worse because the kiss took place on a Sunday) and sentenced to two hours in the pillory.

questions. What was true for astronomy held for biology, as well: every declaration about the world was also a statement about God, who had made the world.

In the 1600s and 1700s, virtually all the major scientists in Europe were devout Christians who shared a deep faith that their mission was to discover God's reasons for designing the world as he had. Questions that seem ludicrous to us—*Did Adam have a navel? Were the lions in the Garden of Eden vegetarians? If lust is a sin and there was no sin in Eden before the fall, how did Adam and Eve have sex?*—struck them as crucial. Since the world was the work of God and the Bible the word of God, it fell to scientists to bring the same reverent scrutiny to both.

The deer in King Charles's royal parks and the birds and beasts in Eden were equally valuable witnesses to the nature of God's creation, and equally real. To dismiss such mysteries as Adam's navel would be as scandalous as if modern scientists were to ignore glaring facts—the eruption of a supposedly dormant volcano or the discovery of fossilized bones from an unknown creature—because they did not know what to make of them.

So it was seen as vitally important to both science and Christianity that Saint Augustine had explained, for instance, what sex looked like in Eden. Just as our hands and feet move under our command, Augustine wrote, in Eden every organ was a "ready servant of the will." Lust did not come into it. In the Garden, one modern historian explains, Adam commanded his penis to rise or fall as needed, "rather like a drawbridge."

Sex in Eden was a sedate affair, performed with what Augustine called "tranquility of mind." There was no question of disturbing the neighbors, even if there had been any neighbors to disturb. And in heaven there was no sex at all. Men and women would have their familiar bodies in heaven, theologians taught, but the women's would have been repurposed. "The female parts, not suited to their old uses, will achieve a new beauty," Augustine wrote, "and this will not arouse the lust of the beholder (for there will be no lust). Rather, it will inspire praises of the wisdom and goodness of God."

Everyone in heaven would be thirty, according to Augustine, the age that marked the peak of bodily perfection. (For those who died younger, God would set the clock forward.) Theologians wrestled with countless similar riddles, endlessly pondering such questions as whether God would restore arms and legs lost in battle or devoured by sharks. The case of cannibalism provoked some of the thorniest debates. *Whose body was whose?* Augustine concluded that "the eaten flesh will be restored by God to the man in whom it first became human flesh. This flesh can be looked upon as a loan taken by the famished man and, like any other borrowed goods, must be returned to the one from whom it was taken."

Those restored and heavenly bodies, though they would not engage in sex, would enjoy a variety of other pleasures. Singing God's praise ranked near the top. "All our activity will consist in singing 'Amen' and 'Alleluia,'" Augustine declared, and he assured his readers that their delight in this entertainment would last forever.* (This was a bold claim, Augustine acknowledged, since heavenly days "have no end in time." *Forever* meant *forever*.)

Other Christian sages focused on different heavenly entertainments, including the opportunity to watch sinners in torment. Heaven came with portholes on hell. "In order that the happiness of the saints may be more delightful to them and that they may render more copious thanks to God for it," wrote Saint Thomas Aquinas, "they are allowed to see perfectly the sufferings of the damned."

Schadenfreude has seldom risen to such heights. Isaac Watts, the seventeenth-century theologian who wrote "Joy to the World" and hundreds of other hymns, explored the theme in poetry: "What bliss will fill the ransomed souls / When they in glory dwell, / To see the sinner as he rolls, / In quenchless flames of hell."

This was a long way from the Song of Songs: "Let him kiss me with the kisses of his mouth: for thy love is better than wine."

---

*To each his own. Mark Twain disagreed with Augustine: "Of all the delights of this world man cares most for sexual intercourse. He will go any length for it—risk fortune, character, reputation, life itself. And what do you think he has done? He has left it out of his heaven!"

FOUR

UNMOORED IN TIME

THE FIRST GREAT ADVANCE IN THE SEX AND BABIES MYSTERY came shortly before 1500, when the squeamish but voraciously curious Leonardo da Vinci took knife in hand and turned his attention to anatomy.

That was new. For two thousand years, to study medicine had been to pore over ancient texts. Medicine was an academic exercise, a debate over exactly what Hippocrates or Galen had decreed long ago. Then, beginning with Leonardo and a few other pioneers, medicine shifted course. The new approach, revolutionary in its daring, was to look with one's own eyes rather than those of a bygone sage.

Most often, those searching eyes belonged to anatomists, who boldly took scalpel in hand and sliced their way past muscle and gristle and inside the human body. The very word "autopsy," a hybrid that combined "to see" and "oneself," enshrined the new ideal. Though we often read that dissection was taboo in Europe all through the Middle Ages, that is not so. Bodies had been cut open long before the Scientific Revolution. What changed was not the fact of the cutting but the motive for it. The new motive was to learn how this complicated living machine worked.

From roughly 1300 to 1500, opening the body after death was common, usually for embalming. In the case of saints, the body was often disassembled so that various bits—hearts, hands, fingers, bones, skulls, vials of blood—could be distributed for the veneration of the faithful. Tourists in Siena today can still gaze upon Saint Catherine's mummified head in an elaborate gold reliquary; her right thumb rests nearby in a smaller shrine. In Padua, a few hours north, the tongue of St. Anthony lies in an ornate shrine of its own.

The ancient world, in contrast, *had* largely banned the cutting up of human bodies. For the Greeks and Romans, to slice the body apart was an insult to the dead and perhaps a ruinous mistake as well, if it meant depriving a beloved relative of a body he was going to need in the afterlife. (Cleopatra supposedly ignored the taboo, perhaps because she was dealing with slaves. According to one eminent historian, she gave orders calling for the execution of pregnant slave girls at fixed intervals after conception, so that she could see how infants developed in the womb. And Nero, seldom outdone even in depravity, murdered his mother and then had her body cut open, at least according to legend, because he was curious "to see the place where he was conceived.")

Dissection was a revolting business. Even Aristotle, a devoted student of the natural world and a man of almost boundless curiosity, acknowledged the "great disgust" we feel when "we see what composes the human species: blood, flesh, bones, veins, and similar parts."

But the ancients had wavered in their views. Over the course of the fourth and third centuries BCE, Greek physicians in Alexandria had carried out hundreds of careful, detailed dissections of human cadavers. That era proved an exception. Neither Hippocrates nor Galen, the two greatest figures in the history of early medicine, ever dissected a human body. (Hippocrates lived around 400 BCE in Greece, and Galen around 150 CE in Rome and elsewhere in the Roman empire.) Both men relied instead on analogies from animal dissections.

Anatomists in Roman days competed for attention, and Galen loved a spotlight. Part carnival barker and part medical lecturer, he performed dissections before swarms of gawking spectators. He favored pigs and

FIGURE 4.1. Nero watches as physicians probe his mother's body.

primates, though on one notable occasion he outdid his rivals by dissecting an elephant. For direct observations of *human* anatomy, Galen had to grab whatever opportunities fortune sent his way. Once he hurried to examine a body that a flood had washed from its grave. The flesh had rotted away, but Galen noted with fascination that the skeleton was still intact.[*]

Sometimes he did not have to wait for floods. In what is now Turkey, Galen's duties included tending to the gladiators who fought for the entertainment of tens of thousands of roaring spectators. Matches ended in death or surrender, with the crowd choosing whether the loser would be spared to fight another day or dispatched with a sword to the throat.

---

[*] He missed another chance when the emperor Marcus Aurelius dragooned him to accompany the Roman army into battle against a host of German tribes. Galen dodged the assignment, only to learn later that the emperor had allowed his physicians to dissect one or more slain "barbarians" from the enemy's ranks. "There can be no doubt that if he had known he would be allowed to dissect a human, he would have braved the perils and discomforts of the campaign," observes one recent biographer.

Gladiators were typically slaves or prisoners of war, but some were free men who had, one historian tells us, "taken an oath agreeing to be burnt, chained, beaten, and killed with an iron weapon." In return, they had a (slim) chance at fame, glory, and riches. They suffered gaping wounds inflicted by sword and dagger. These provided Galen with tests for his surgical skills. Better yet, he noted contentedly, they offered "windows into the body."

The early church insisted that it was sinful to peep through any such window. Humankind's task was to rise above the body, not to immerse itself in the contemplation of its muck and fluids. "It is far more excellent to know that the flesh will rise again and will live for evermore," wrote Saint Augustine, around the year 400 CE, "than anything that scientific men have been able to discover in it by careful examination."

Since God had hidden the body's secrets from prying eyes, Augustine argued, it was impious to try to subvert his intentions. The anatomists' "cruel zeal for science" had led them astray. Curiosity was a sin, not a virtue, and in fact a deadly sin. Augustine railed against it with fury. To study nature or even the inanimate world, Augustine wrote, was to indulge "the lust of the eyes." This was perversion. A person might just as well gawk at sideshow freaks or stop at a roadside accident to stare at "a mangled carcass." We picture scientists as explorers of the unknown; our forebears saw them as peeping Toms.

Augustine's denunciation of curiosity prevailed for a thousand years. To ask questions was to flirt with skepticism, and skepticism was but a step from heresy. Who were human beings, creatures made of clay, to question their Creator? Faith was the essential virtue, pride the great danger. "Knowledge puffeth up," Paul declared in his first letter to the Corinthians, and humankind had a duty to bear that rebuke constantly in mind.

So ran the standard argument, and for century after century conventional thinkers thumped out indignant variations on the same theme. Insight was best found in books, preferably venerable ones. To look for oneself was a mark not of independence but of foolish impudence, as if a layman proposed that he could build a sturdier ship than a professional. The new scientific credo—*think for yourself*—threatened to turn the world upside down. "If the wisest men in the world tell them that they see it or know it; if the workers of miracles, Christ and his apostles, tell them that they see it; if God himself tells them that He sees it," one theologian thundered in 1665, "yet all this does not satisfy them unless they may see it themselves."

Today the insistence that we defer to authority sounds strange and wrongheaded. Why take someone else's word for what we could test ourselves? How can there ever be progress if we constantly look to the past for guidance? But we live in a world shaped by an intellectual revolution. That revolution was a scientific one, and its heroes were not generals but intellectuals. Their victory was so complete that we take it for granted, to the point that we scarcely remember that we've built our homes on what was once a battleground.

We assume, for instance, that "new" is a word of praise, almost synonymous with "improved." But that idea itself is new. Our forebears would have found it bewildering. Since all God's works were perfect, nearly any change was suspect. We see old ideas as musty and perhaps irrelevant. They saw them as time-tested and stamped with the endorsements of unimpeachable authorities.

The nearly unquestioned view in the 1500s and 1600s was that the world had declined since Eden, not just spiritually but intellectually as well. The Fall brought not only sin and death into the world but every sort of bad thing, from tiny to momentous. After the Fall, snakes turned venomous, roses sprouted thorns, and human insight dimmed.

Nearly everyone believed that everything worth knowing had been found long ago. History did not progress but repeated the same cycles, endlessly. Only the names changed. "Again will Achilles go to Troy, rites and religions be reborn, human history repeat itself," one scholar wrote in 1616. "Nothing exists today that did not exist long ago. What has been, shall be."

The very words that early modern scientists used to describe their quest reveal their reverence for the past. "To 'dis-cover' was to pull away the covering cloth, disclosing what may have been hidden, over-looked, or lost, but that was in any case already there," explains the historian Darrin McMahon. "To 'invent,' similarly, was to access that *inventory* of knowledge long ago assembled and put into place: an invention was just a dis-covery, a recovery of an object forgotten."

L EONARDO DA VINCI WAS ONE OF THE EARLIEST OF THESE ASTON-ishing, ambivalent revolutionaries, split in his own mind between a past he revered and a future he only dimly glimpsed. Starting in about 1490, he set down a series of extraordinary anatomical drawings in his notebooks.

Sexual anatomy was a compelling part of the story Leonardo wanted to tell, but only a part. Characteristically, he set out to explore *every-thing*. His precise, beautiful drawings depicted almost all the muscle groups in the body, and bones and blood vessels and organs, as well as male and female genitals. "Virtually every drawing is the finest depiction of a particular structure to that date," one scholar notes, "and in some cases for several centuries to come."

Leonardo's sexual studies broke new ground, and at the same time they highlighted how little was known for sure. Even worse, much of what was "known" would later prove false. This is no reflection on Leonardo, who had one of the keenest eyes in world history. The human body is astonishingly complex; it is no wonder that its first explorers occasionally ventured off course. We think of Leonardo as modern, with his helicopters and submarines, but he lived in an age when science and medicine had barely advanced beyond ancient doctrines.

It was a coincidence—although a coincidence that reflected a change in Europe's intellectual climate that was destined to shake the world—that Leonardo made his first, revelatory drawings of the human body at almost precisely the same time that Columbus set out to explore a new world of his own. (Columbus's discoveries—a continent that the ancients had never imagined, teeming with animals and plants they had not heard of—helped undermine the authority of Aristotle and other venerable writers. Every ordinary sailor on Columbus's voyages, the great English scientist Robert Boyle observed, had seen "an hundred things that they could never have learned by Aristotle's philosophy.")

In the early 1490s, Leonardo was about forty and had already mastered more fields than he could list. A decade earlier he had written a letter to the Duke of Milan seeking an appointment as a military engineer. Leonardo described his ideas for new sorts of cannons, tanks, and catapults. Then he added a kind of P.S.: "Also I can execute sculpture in marble, bronze, and clay. Likewise in painting, I can do everything possible as well as any other, whosoever he may be."*

Engineer, inventor, musician, painter, sculptor, Leonardo certainly conveyed the impression that he could do "everything possible." His appearance was as eye-catching as his résumé. A man of "supernatural" beauty, in the words of one of his earliest biographers, Leonardo was tall and muscular, with long brown hair. He dressed almost entirely in pinks and purples, in satin and silk and velvet, his cloaks, caps, and leggings an exuberant contrast with the somber outfits of other men, and he traveled with a bevy of young, beautiful admirers.

One of the most famous of his anatomical drawings, from around 1492, is a cutaway drawing of a man and woman having sex. Scattered on other notebook pages are drawings of male and female genitals, some on their own, some locked together, like doodles from the world's most talented schoolboy. As if in testimony to Leonardo's varied interests, the genitals share the pages with carefully rendered drawings of cranes, pulleys, and levers. Leonardo's mirror-writing spills across the paper. (The writing, contrary to legend, is not especially hard to decode and would have been a poor way to record secrets. Apparently Leonardo took up backward writing as a sort of party trick when he was young, and then got in the habit.)

The most detailed of the sex sketches looks as if Leonardo has somehow, with X-ray vision, drawn a couple making love. That could

---

*The duke would later commission Leonardo's *Last Supper.*

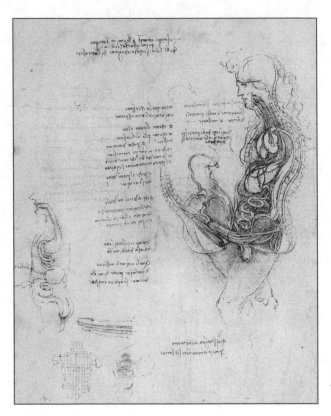

FIGURE 4.2. Leonardo drawing, ca. 1492.

scarcely be, of course, but the drawing is not even a depiction of what you would see if you could freeze the action, take a scalpel, cut open a man and woman, and record the scene. The drawing shows not what Leonardo saw when he peeked inside the human body—he would not carry out his first human dissections, except in hit-or-miss fashion, until about a decade *after* this sketch—but what outdated texts told him he would see. His drawing serves less as a depiction of the body at play than as a guide to the medieval mind at work.

Leonardo's depiction of the penis, for example, is a hodgepodge of ancient Greek theories and medieval guesswork. The drawing shows two distinct channels within the penis, though in fact there is only one. As Leonardo drew things, the lower channel carries urine while the upper carries semen and connects with the spinal column. (The role of the testicles in all this was not quite clear.) The spinal connection reflected a Greek belief that, in the words of one ancient writer, "sperm is a drop of brain."

Leonardo's transparent woman has design peculiarities of her own. For a start, she lacks ovaries. As if to make up for that oversight, she has a mysterious tube running from uterus to nipple. This nonexistent pathway, like the imaginary one supposedly running between penis and spinal column, reflected medical dogma. In the case of men, the notion was that sperm was refined from the blood and then made its way to brain, spine, and penis. In the case of women, the refining process transformed menstrual blood into mother's milk. (The theory, dreamed up by the Greeks, was an attempt to explain why pregnant women and new mothers did not menstruate.)

It was not all guesswork and misinformation. Leonardo did explain correctly, for the first time in history, how what he called the "soft and feeble" penis grows erect. At last, one of our detectives had happened on an actual clue! The conventional explanation had always been that the erect penis was inflated with air, like a beach ball or one of those New Year's Eve noisemakers that uncurl to full length when you puff into it. (This belief was so widespread that, as late as 1671, one well-regarded sex and childbirth manual noted that "windy" foods—"beans and Pease and the like" —"will make the Yard stand.")

Leonardo contemptuously rejected such notions. "If an adversary says that such a large quantity of flesh has grown through wind causing the enlargement and hardness, as in a ball with which one plays, such wind gives neither weight nor density but makes flesh light and rarified." But the penis was heavy and clunky, not light and airy. Certainly this odd appendage did not drift aloft like a flimsy and insubstantial balloon.

Leonardo took up this mystery on several occasions. In the margin of a page that had nothing whatever to do with the genitals—the sheet contains a dazzlingly rendered drawing of the muscles in the lower leg and the tendons in the foot—he jotted down another objection to the standard view. "There would have to be a great wind

FIGURE 4.3. Hanged man, by Leonardo.

to enlarge and elongate the penis, and make it as dense as wood. . . . Even if the whole body were full of air, there would not be sufficient."

He had reached a different conclusion, perhaps as early as December 1479, when he witnessed the hanging of a Florentine assassin. Bernardo Baroncelli had thrust a knife into the chest of Giuliano de Medici, the brother of the city's ruler, Lorenzo the Magnificent. Now he dangled from a rope, hands tied behind his back, while a boisterous crowd looked on. Leonardo sketched the dead man in a few swift lines. Alongside the drawing he jotted down several dispassionate notes: "a tan-colored small cap," "a blue coat lined with fur of foxes' breast," "black hose."

Years later he recorded another observation, perhaps from this early scene or perhaps from another execution. The dead man had an erection. "Many die thus, especially those hanged," Leonardo wrote. Why were the hanged hung? Leonardo sorted that out. Postmortem dissections of newly executed criminals resolved the mystery. "I have seen the anatomy [of these penises]," Leonardo wrote in a notebook, "all of them having great density and hardness, and being quite filled by a large quantity of blood." Blood, not air. Case closed.

This was a genuine advance, but Leonardo quickly passed onto other matters. Perhaps so straightforward a bit of engineering as male hydraulics held less intrigue than did flying machines and diving suits.

Nor did the contemplation of sex speak to Leonardo's aesthetic side. "The act of coitus and the parts employed therein are so repulsive," he wrote in a notebook, "that if it were not for the beauty of the faces and the adornments of the actors and frenetic state of mind, nature would lose the human species."*

Leonardo's professed disgust with humankind's sexual apparatus may have been a bit one-sided. At times he described the penis with fond indulgence, as if he were speaking of an eager but exasperating employee. "It remains obstinate and goes its own way," he wrote. " . . . Often a man is asleep and it is awake, and many times a man is awake and it is asleep. Many times a man wants to use it, and it does not want to; many times it wants to and a man forbids it."

As Leonardo described things, this was an organ that seemed more like a living animal. Moreover, "it appears that this animal often has a soul and intellect separate from a man; and it appears that a man who is ashamed to name or show it is in the wrong, always being anxious to cover up and hide what he ought to adorn and show with solemnity."

FOR ANATOMISTS IN LEONARDO'S DAY, THE FIRST PROBLEM WAS finding a body. Newly executed criminals were the usual subjects, but there were not enough of them to go around. Animals were easier to come by. Leonardo dissected whole menageries. Around 1490 he made a meticulous drawing of the muscles and tendons in a bear's paw, for instance, as well as anatomical drawings of dogs and monkeys. At about this time, he managed to acquire a human leg, which he dissected and drew. He found a human skull, too, sawed it open, and produced a careful series of sketches.

More than a decade would pass before Leonardo finally had a reliable supply of human bodies to dissect, from hospital patients who had died and whose bodies had gone unclaimed. On one winter's day in the early 1500s, he sat at the bedside of an ancient, frail man in a hospital

---

*The comment turns up, out of the blue, on a page of spectacular drawings of the hand that show in detail how its muscles, bones, and tendons interact.

for the poor in Florence. "This old man, a few hours before his death, told me that he was over a hundred years old, and that he felt nothing wrong with his body other than weakness. And thus, while sitting on a bed in the hospital of Santa Maria Nuova in Florence, without any movement or other sign of any mishap, he passed from this life."

In the very next sentence, Leonardo noted coolly that "I dissected him to see the cause of so sweet a death." (He found two problems, hardening of the arteries and cirrhosis of the liver. Neither had ever been described before.) In the same entry in his notebook, and in the same matter-of-fact tone, Leonardo recorded "the dissection of a child of two years, in which I found everything contrary to that of the old man." Over the course of the next several years, now with both executed criminals and hospital patients to study, Leonardo plunged ahead. He would soon boast that he had dissected "more than thirty" corpses.

At the best of times, dissection is a messy business. The 1500s were not the best of times. In an era before refrigeration and before the development of embalming techniques, this was, in the words of Leonardo's first biographer, "inhuman and disgusting work." A note that Leonardo wrote to himself before a trip in 1510 gives some hint of what his dissections involved and, incidentally, of his difficulty in confining himself to one project at a time. He reminded himself to pack forceps, a bone saw, and a scalpel (as well as "boots, stockings, comb, towel, shirts"), and he jotted down a few specific tasks. "Get hold of a skull." For best results, "break the jaw from the side." "Describe the tongue of the woodpecker and the jaw of a crocodile."

Unlike Leonardo's sex drawings, which he made early in his career and which illustrated theories that other men had proposed, Leonardo's later anatomical sketches rely on observations that he made himself. Presumably he took notes while he worked, but working in the dissecting room would have been more like drawing at a crime scene than in an artist's studio. The grime-smeared first drafts vanished long ago. We have only the finished drawings, crowded but so clean and precise that we can hardly imagine Leonardo standing next to his wet, slick subjects and sketching by the light of a flickering candle.

The very years in which Leonardo spent his days and nights manipulating lifeless flesh were also those when he painted images whose beauty would make them immortal. In Leonardo's mind, at least, the anatomical drawings and the soft, subtle portraits bore a close connection. At the same time he was painting the *Mona Lisa*, for instance, Leonardo was cutting open the faces of corpses and dissecting the muscles of the mouth and lips, to sort out the secrets of the smile. "These I intend to describe and illustrate in full," he wrote, "proving these movements by means of my mathematical principles."

Nor was that all. "It is noteworthy," writes Kenneth Keele, a renowned Leonardo scholar and a physician as well, "that at about the same time as Leonardo was painting the *Mona Lisa* he was making his dissections of the pregnant uterus." That dissection yielded perhaps the most famous of all Leonardo's anatomical drawings. The sketch, based on Leonardo's dissection of a woman who died when she was about five months pregnant, depicts a curled-up fetus in profile, with all ten toes

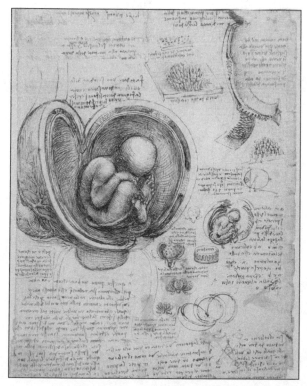

FIGURE 4.4. Leonardo drawing, ca. 1510.

carefully rendered, and a big, domed head and a tiny ear and a winding umbilical cord.

We peek inside the opened-up womb as if at a stage set where, a moment before, the curtains have been pulled back. Keele believes that, at just this period, Leonardo's various interests converged. "My interpretation of the smile of Mona Lisa," he writes, "is that it subtly expresses the secret, which she has successfully kept for so long, that she is pregnant."

The stunning ink and red chalk drawing of an infant in the womb was one of hundreds of anatomical sketches that Leonardo made in a few intense years around 1510. The "notebooks" that include these drawings were really loose pages that were assembled years later. If Leonardo had a particular order or arrangement in mind, scholars have never been able to guess it. Most sheets are covered on both sides with beautifully rendered drawings and careful notes.

Leonardo described his accomplishment proudly. "You who say it would be better to watch an anatomical demonstration than to see these drawings would be right," he challenged, "if it were possible to see all the things the drawings demonstrate in the dissection of a single body." But no single dissection could show the different levels of organization that Leonardo drew, with his cutaways, close-ups, and multiple vantage points.

After Leonardo's death in 1519, his anatomical sketches and notes disappeared from view. (Leonardo was a master of the art of procrastination, along with every other art.) He had enormous difficulty finishing projects, and though he had grand ambitions for a book on anatomy, he never published one, or anything at all, during his lifetime. He bequeathed countless manuscripts and artworks, including his anatomical drawings, to a young painter and protégé named Francesco Melzi. Over the course of fifty years, Melzi struggled to impose some order on this ingenious shambles. He paid little heed to the anatomical work. In time Melzi's son inherited his father's Leonardo collection and sold it.

The anatomical sketches passed to an Italian sculptor. Eventually they wound their way to England. They landed at a suitably grand location, Charles II's royal library. Even so, they lay neglected through the centuries, not only unpublished but seldom even glanced at. Not until 1796, almost three hundred years after Leonardo's death, would the world see what he had found so long before.

In the meantime, other anatomists had joined the quest. They did not have Leonardo's discoveries to learn from, and they did not know that, when it came to the mysteries of sex and conception, this supremely confident thinker had struck an uncharacteristically modest note. "There is much that is mysterious," Leonardo conceded briskly, and then he hurried on to more congenial topics.

FIVE

"DOUBLE, DOUBLE TOIL AND TROUBLE"

LONG AFTER LEONARDO'S DEATH, ANATOMISTS CONTINUED TO have trouble finding human subjects to dissect. For centuries they relied on newly executed criminals for most of their raw material. This was mainly a matter of practicality—doctors needed bodies to study, and criminals, many of them isolated and friendless, had no say about whether they wanted their last act to play out on a dissecting table. Partly the story was even darker. Dissection was a fearful prospect, widely seen as too harsh an end to the life of an ordinary man or woman but a fitting finale to a criminal's career of sin. (In England, in 1752, Parliament would make this reasoning official. The newly passed Murder Act decreed that a condemned murderer be dissected as well as hanged, in order "that some further terror and peculiar Mark of Infamy be added to the Punishment of Death.")*

Rembrandt's *Anatomy Lesson of Dr. Joan Deyman*, from 1656, captures the ghastliness of dissections in the 1600s and 1700s. The naked

---

*Prisoners awaiting death had to fend off anatomists, or their agents, who tried to cajole them into bartering their soon-to-be corpses for money. Some condemned men arranged deals. They used the money to ease their last days in prison or for new clothes to wear to the gallows.

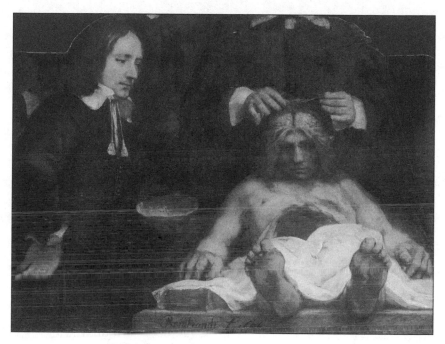

FIGURE 5.1. Rembrandt, *Anatomy Lesson.*

man at the center of the painting was a thief known as Black Jan, who had been executed by hanging. The prisoner's exposed brain is the focus of the day's lesson. His intestines and stomach have already been removed, as was customary, because they spoiled quickly. Jan's dirty feet jut out toward the viewer; his head sits askew on his chest because his neck was broken on the gallows. High above the stage in Amsterdam's Anatomy Theatre (but not depicted in Rembrandt's close-up view), a few words inscribed in golden letters offered a moral: "Evil men, who did harm when alive, do good after their deaths."

These macabre scenes played out a bit like seventeenth-century horror films, but also as something more edifying, perhaps even uplifting. Here was a chance to see what physicians of the day called "secrets of nature revealed by God." In Amsterdam, Bologna, London, Padua, Paris, and every other city that aspired to the first rank in medicine, students and curious laypersons shoved their way into anatomical theaters like the one Rembrandt depicted. Hundreds of spectators jammed together, in steeply rising tiers so that everyone could see the corpse on the stage below, displayed on a rotating, wheeled table.

Dissections of female bodies drew the biggest crowds, partly because female subjects were rare. A pregnant woman drew best of all, since the anatomist could take up such topics as the mysterious womb. Many of these women had been hanged as thieves or prostitutes. A few had died in childbirth, unmarried and abandoned.

Public dissections were performances carried out with elaborate ceremony. Rules stipulated that the ground be covered with a mat, so that "Mr. Doctor be made not to take cold upon his feet" and spelled out exactly how many aprons, knives, candles, and other tools the anatomist would need. In most venues a chandelier crowded with scented candles hung above the stage.

Like audiences at a play, noblemen in elegant dress craned to see, as did tradesmen in their work apparel, tourists making the rounds, courting couples eager for a bit of a frisson on their night out, and even children who had managed to sneak their way past the doorkeepers. A flutist provided tasteful accompaniment. In Amsterdam, rules forbade laughing and talking, as well as stealing hearts or kidneys or livers from the victim's body as they made their rounds through the audience.

Ticket prices varied from nothing at all in Padua "in order that everyone may come" to enjoy the spectacle to the equivalent of a few dollars in Holland. The surgeons' guild in Amsterdam customarily used some of its profits for a lavish postdissection banquet (and a generous tip for the hangman), with wine and tobacco and a parade to cap the evening.*

Dissections took place in winter; summer's heat would have made it nearly unbearable to delve into an open body. Even so, the whole procedure was a race against decay. In the interest of speed, different bodies were used to show different internal structures. One corpse might serve for a lecture on muscles, a second for bones, and a third

---

*Dutch archival records show that Dr. Deyman, the surgeon in Rembrandt's painting, earned six silver spoons for his labors, the equivalent of several hundred dollars in today's money.

for internal organs, in much the same way that anatomy books would one day use drawings on different layers of transparent plastic to depict heart and lungs, and bones, and nerves.

N EVER A BASHFUL GROUP, ANATOMISTS CONFRONTED THE SHORT-age of bodies with ingenuity and gusto. The first and most esteemed of all modern anatomists, a brash young Belgian named Andreas Vesalius, had showed the way in the mid-1500s. "First you must obtain a corpse from somewhere," he wrote. "What sort does not matter, though one emaciated by disease is much the best."

That passage occurs in Vesalius's masterpiece, *On the Fabric of the Human Body.* The book appeared in the momentous year 1543, which

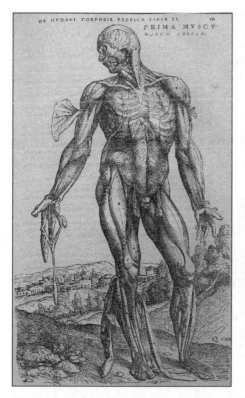 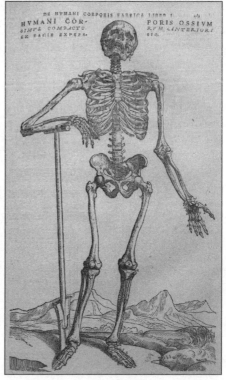

FIGURE 5.2. Two images from Vesalius's masterpiece, *On the Fabric of the Human Body.*

saw the publication of two of the most important books in Western history. One was Copernicus's *On the Revolutions of the Celestial Spheres*, which proclaimed that the Earth travels around the sun and not vice versa. The second was Vesalius's atlas of human anatomy, one of the most significant and most beautiful books ever published.

That beauty rose from hideous roots. Vesalius first ventured into anatomy, he wrote, on a day in 1536 when he had the good fortune to find a corpse dangling from a gibbet. In his telling, his excursion in search of a body was a lark, the kind of boys-will-be-boys prank that Tom Sawyer might have got up to if his brand of mischief had run less to tricking his friends into painting a fence and more toward tugging bones free from a dried-up corpse.

Vesalius had set out with a friend "in the hope of seeing some bones." In the sixteenth century, this was no great challenge. "We went to the place where, to the great advantage of students, all who have suffered the death penalty are displayed by the public highway for the benefit of the rustics." There the two adventurers found a body that had been hanging outdoors for a year. The victim had evidently been burned alive, as punishment for some forgotten transgression. Vesalius described the scene lightheartedly. The dead man had been "as it were toasted over a straw fire, and, tied to his stake, he had provided the birds with such a tasty meal that the bones were completely bare."

Vesalius delighted in his unexpected good fortune. Typically, "despite a popular image to the contrary, the birds normally peck away nothing but the eyes, because the skin is so thick, and as the skin remains intact, the bones decay inside and are quite useless for teaching purposes." But not this time! With his friend's help, Vesalius clambered up the stake and wrestled a thighbone away from the hip. Then he yanked off a shoulder blade, and arms, and a hand. He sneaked his treasures home and returned for more, in the dead of night. "So keen and eager was I to obtain these bones that I did not flinch from going at midnight amongst all those corpses and pulling down what

I wanted." In the end, Vesalius managed to find nearly everything he needed (except one hand and one foot, which he pulled from another body). He treated and cleaned the bones, strung them together, and then proudly unveiled a full skeleton.

Doctors in ages past, Vesalius scolded, had prepared bodies for study in ways that were "troublesome, dirty, and difficult." Scornful of those approaches and thrilled with his own innovations, Vesalius happily shared his techniques. Everyone could explore these mysteries. All they had to do was follow a few simple steps.

Utterly unsqueamish, Vesalius wrote as if he were teaching a cooking class for Macbeth's witches. "Having thus placed the bones in the cauldron," he instructed his readers, "fill it up with enough water to cover the bones completely." He dotted his text with tips and warnings. "As usual in boiling, the scum should be carefully removed," he advised, and he offered the useful reminder that "the object of the seething is to enable the bones to be scraped clean with a knife, as in meat for eating."

IT IS EASY TO MISUNDERSTAND VESALIUS AND TO CONCLUDE, FROM the light, self-satisfied tone in which he discussed his dark adventures, that he was a brutal man. We should be wary of such judgments. The past is indeed a foreign country and, by our standards, an astonishingly violent one. The most routine sights in sixteenth-century France or seventeenth-century England would leave us reeling in horror. For years on end, for instance, the severed heads of executed criminals stared out from spikes mounted on London Bridge and elsewhere around the city. The dried, leathery heads drew scarcely any notice.

Sightseers in Shakespeare's London who tired of puppet shows and bear baitings might choose instead to visit an asylum; the ravings of lunatics made excellent entertainment. So did the moans of a prisoner locked in the pillory, as the wretch turned this way and that to dodge the rotten tomatoes and dead cats hurled at him.

The modern world prefers to hide its cruelties. The world of our forebears had no such qualms. Hangings drew boisterous crowds that sometimes numbered in the tens of thousands. Spectators gulped their drinks and devoured their lunches while cheering for those miscreants who managed a few jaunty words when the hangman fit a noose around their neck.

The taste of the educated and refined was no daintier than that of the rabble. On the evening of October 13, 1660, Samuel Pepys wrote in his diary that he had gone "to see Major-General Harrison hanged, drawn, and quartered," which is to say, hanged but taken alive from the gallows and then disemboweled, as the victim himself watched. "Which was done there," Pepys wrote jokily, "he looking as cheerful as any man could do in that condition. He was presently cut down, and his head and heart shown to the people, at which there was great shouts of joy." A few sentences later, in the same diary entry, Pepys tells us that he had oysters for dinner.

Vesalius would happily have joined Pepys at his meal. The chapters of his masterpiece begin with elaborate, frank illustrations. Book I's drawing depicts the chubby, adorable toddlers known as *putti* boiling a body in a kettle, to prepare a skeleton. Book VII opens with a drawing of *putti* robbing a grave.[*]

A S THE FIELD OF ANATOMY GREW BETTER ESTABLISHED, THE hunt for bodies intensified. Taboos against dissection ruled out most "respectable" bodies. That called for ingenuity. Laws spelling out who could be dissected varied from country to country. Italy preferred criminals but in a pinch made do with "Jews or other infidels." England seemed less likely to run short of material, because in theory a long list of offenses—some as minor as cattle rustling and even shoplifting—could draw a death sentence. In practice that seldom happened, and

---

[*]Fear of dissection had a long history. Shakespeare died in 1616. His epitaph reads: "Good friend, for Jesus sake forebear / To dig the dust enclosed here. / Blest be ye man that spares these stones / And cursed be he that moves my bones."

the supply of bodies from executed criminals could not keep up with the demand. (William Harvey's college at Cambridge had a contract assuring it the bodies of two criminals a year, freshly executed and free of charge.)

For centuries, scarcely any space separated the study of anatomy from the work of the hangman. In Britain, in 1752, that gap closed completely. A new but poorly thought-out law made it illegal to dissect anyone except a criminal fresh from the gallows. That made for fewer bodies than ever available for study. As a result, body snatching became a high-paying criminal specialty.

So-called resurrection men crept into cemeteries in the dark and sneaked their way to a fresh grave. Wielding wooden trowels—so that the *clink!* of a metal shovel against a rock or the coffin didn't give the game away—the grave robbers dug a shaft straight down to the head of the coffin. For speed's sake, they left the coffin undisturbed along most of its length. Next the resurrection men tied a rope around the coffin's exposed end and stuffed a canvas sheet into the freshly dug hole, to muffle any sounds.

Then a sharp yank upward on the rope! Wedged into the ground, the coffin cracked open. The grave robbers grabbed their prize by the armpits and pulled it free. They tidied the site so it would not catch a watchman's eye and raced off. Elapsed time: under an hour.*

A fresh body delivered to an anatomist's door might bring as much money as a workman could earn in six months. Newspaper headlines warned of "Church Yard Pirates." To thwart body snatchers, coffins were secured with metal bands or placed inside iron frames.† Not content with such defensive measures, some people booby-trapped

---

*The poet Thomas Hood wrote a grimly comic poem in 1826 called "Mary's Ghost," where Mary tells her sad story to her beloved William: "I thought the last of all my cares / Would end with my last minute, / But when I went to my last home / I didn't stay long in it. / The body-snatchers, they have come / And made a snatch at me. / It's very hard that kind of men / Won't let a body be."

†Body snatching was against the law, but the law was a muddle. Since a body was not property, it did not belong to anyone. Stealing a shroud carried higher penalties than stealing the body within it.

their relatives' coffins with gunpowder, rigged to explode if anyone tried to break in.

For decades, everyone knew about anatomy's dark side, but no one confronted it. Finally, in Scotland in 1828, matters took an especially hideous turn. Two Irish immigrants living in Edinburgh decided that grave robbing was too cumbersome a way to obtain corpses. Instead, over the course of twelve months, William Burke and William Hare murdered sixteen people and sold their bodies to a celebrated anatomist named Robert Knox, at the University of Edinburgh. Murder was an ancient crime, wrote one horrified journalist, but a murder like this "was never, we believe, heard of before."

Burke and Hare (with the help of their wives) coaxed their victims to drink too much. Then they suffocated them and delivered the bodies to Knox. The pair were no masterminds, but the police were scarcely more capable, and they could not put together a strong case. (They had only one body, the others having already been dissected, and they could not prove that the dead woman they found had been murdered, because suffocation left no marks.)

In the end, Burke was found guilty and sentenced to death. (Hare was let off, in return for his testimony against his partner.) Dr. Knox claimed not to have known anything whatever about murder. He did not maintain that he had dissected only criminals who had been executed, as the law required, but he insisted that he procured his subjects in the most above-board fashion. His assistants "watched the low lodging houses" to see who was ailing, Knox explained, and then purchased the bodies of newly dead paupers from their relatives. Knox was never charged and went on to practice medicine for another three decades.

Burke was hanged before a crowd of thirty thousand gleeful, shouting, hat-waving spectators on a cold and rainy January morning in 1829. The next day he was dissected (but not by Knox) at the University of Edinburgh. His skeleton is still on display at the university's Anatomy Museum. As perhaps a more significant legacy of the murders, Parlia-

FIGURE 5.3. The skeleton of the murderer William Burke, at Edinburgh Medical School. The judge ordered him hanged and then dissected and put on public view, "in order that posterity may keep in remembrance of your atrocious crimes."

ment passed the Anatomy Act in 1832. The new law put an end to the practice of dissecting criminals who had been executed. Instead, it gave medical schools permission—despite the fearful protests of many of the poor—to dissect bodies that had gone unclaimed at death.

All these difficulties—finding bodies to study; fighting down the horror of cutting up corpses in cold, dank rooms; deducing the workings of the living body from the structures of a dead one—plagued every early anatomist.

These were formidable obstacles, and in practice they proved even more daunting than anyone had anticipated. For those looking to understand the mysteries of sex and conception in particular, the challenge was greater still. Vesalius, a man whose confidence spilled over and ran onto the ground, had rushed eagerly ahead. First he'd pored over a stack of pages that "reached from heaven to earth," to learn what medical men before him had discovered about the physiology of sex.

Alas, his reading explained nothing. Vesalius found himself "storm-tossed upon a mighty sea of opinions and theories." He had plunged ahead with his own dissections, but in the end he conceded that although he had uncovered new facts, he had left the central mystery untouched.

Such frustration was all but inevitable. For Vesalius and the other venturesome explorers in medicine's early days, the deepest problems were conceptual rather than logistical. Confront an unfamiliar sight, and many times you cannot grasp what you are looking at. Take Leonardo. Genius though he was, Leonardo had studied the heart with enormous care without ever recognizing that it is a pump or that blood circulates around the body. Historians have struggled to explain why not.*

They note, for instance, that Leonardo's dislike of vivisection meant that, unlike Harvey, he never saw a heart in action. More important, perhaps, was a different sight that Leonardo missed. By Harvey's day firefighters had begun using powerful pumps that shot out jets of water. Leonardo, born a century earlier, had no such experience to draw on.

Seeing is not the same as understanding. Until you have broken the code, a printed word is not "eagle" or "moonlight" but just an inky squiggle. A hunter's bow is not a weapon but only a bit of wood and cord. An opened-up human body is not an intricate machine but a dark, crowded morass.

The modern-day doctor and writer Jonathan Miller has described the bewilderment of medical students, today, when they first encounter a body's interior. Their immersion in textbook drawings and plastic models helps only so much. "The unsuspecting student plunges into the laboratory carcass expecting to find these neat arrangements repeated in nature," Miller writes, "and the blurred confusion which he actually meets often produces a sense of despair. The heart is not

---

*Butchers had always known that when you cut an animal's throat, blood jets out in a rush. (Soldiers and criminals knew about gushing blood, too.) Before William Harvey, no one cited that familiar observation as proving that blood races through the body. All that it showed, anatomists believed, was what happened in one special, violent case, not in the everyday course of life.

nearly so clearly distinguished from its vessels as the textbook implies, and at first sight the vessels are practically indistinguishable from one another."

And those frustrated students are only trying to confirm what others have already found. Leonardo and Vesalius and Harvey were out to make discoveries, an immensely harder challenge.

SIX

DOOR A OR DOOR B?

F OR THREE-QUARTERS OF A CENTURY AFTER VESALIUS'S DEATH, the *Where do babies come from?* mystery languished in a cold-case file. Then, in the early 1600s, William Harvey took it up. He had two motives. One was ambition. Harvey was the leading anatomist of his day. To solve the greatest anatomical riddle of all, and one that had stymied all his predecessors, would cap a daunting career. The second reason was more important: William Harvey had a new idea.

Harvey set up the mystery as if he had grabbed hold of a telescope—a brand-new invention in his day—and used it to look back in time rather than out in space. "A Man was first a Boy," he wrote. " . . . Before he was a Boy, he was an Infant; and before an Infant, an Embryo." And before that? Harvey had an answer. Hippocrates and Aristotle and Galen had proposed answers, too. For fifteen centuries afterward, everyone had parroted them. They'd all had it wrong, Harvey declared. He would set matters straight.

H ARVEY WAS HIGH-STRUNG, SELF-ASSURED, FAST-MOVING, AND fast-thinking. He scarcely slowed even at night. When he ought to have been sleeping, he found himself awake against his will, and he passed the night hours pacing endless, restless circles in his bedroom. In the anatomy theater, though, he performed with mastery and flair. He strutted around the stage, circling the body laid out before him, with his long, white sleeves flapping as he drew his audience's attention to particular body parts with a wand made of whalebone and tipped in silver.

A jangle of contradictions, Harvey was a bluff countryman who retained his taste for plain food and plain speech but was friend and companion to kings. He was jumpy and restless but cool in circumstances that rattled everyone else. (During the battle of Edgehill, in the English Civil War, Harvey was entrusted with watching over the king's two young sons. At one fraught moment, he curled up under a hedge with the boys and took out a book. "But he had not read very long," a friend wrote afterward, "before a bullet of a great gun grazed on the ground near him, which made him remove his station.") He was a staunch conservative but, when it came to science, a revolutionary.

Most important, he was a medieval man at home in a spirit-haunted world and, at the same time, a modern scientist who helped topple that world. Witches still flew through the night in the 1600s, for instance, and Harvey was once sent (by Charles I) to examine a suspicious old woman. Everyone knew about witches. They knew about spells and potions; they knew that witches rubbed their bodies with the fat of murdered babies so they could slither through tiny cracks into their victim's homes. They knew, too, that Satan provided witches with animal companions who helped them in their dark schemes.

Harvey asked to see this woman's "familiar," which turned out to be not a black cat but a toad. He invented an errand and sent the woman on her way. "His tongs were ready in his hand," an acquaintance of Harvey's recalled, "and he caught up the toad in them. His dissecting knife was ready also. . . . He examined the toad's entrails, heart, and lungs, and it no ways differed from other toads, of which he had dissected many. Ergo it was a plain natural toad."

Carnifex Maiestatis Regis Angliæ

FIGURE 6.1. Harvey was physician to King Charles I, who encouraged his scientific investigations. England fell into civil war in the 1640s, and the king was beheaded in 1649. Charles in good times (*left*) and bad (*right*).

So much for the black arts. The belief in witches dated back to ancient times. The faith in experiments was brand-new. Harvey and his fellow scientists, who believed in alchemy and unicorns and astrology, managed somehow to navigate a world that contained both.

Over a span of many years, Harvey would devote all his intellectual powers to unraveling the riddle of sex. Curiously, he found the whole business off-putting. Sex, "which is itself loathsome," did not inspire him to flights of poetry. But if Harvey did not have much of a lyrical streak, he did have a host of strengths, notably a gift for devising decisive experiments, persistence verging on obstinacy, and a skillful way with a scalpel.

He had unveiled his new picture of the heart in 1628, in a skinny book called *On the Circulation of the Blood*. The title proclaims the breakthrough. Long before Harvey, it had been known that blood moved within the body, but the word "circulation," if it was used at all, referred to a slow, haphazard drift, like the circulation of air in a house.

Harvey proved that the truth was far different. He showed conclusively the falsity of the traditional belief that the body contained two distinct kinds of blood, which sloshed their way along separate pathways and nourished different organs. On the contrary, the same blood made circuit after vigorous circuit around the entire body, nourishing as it travelled, and propelled by the muscular heart.*

Harvey's notion of a quick, purposeful round-trip was startling and new; the description of the heart as a pump was revolutionary. This was not simply a new explanation but a new *kind* of explanation—a mechanical account of what had long been literally a sacred heart. That lofty organ was suddenly brought to earth, transformed not merely into a machine but, jarringly, into a machine made of slick, dark meat.

Around 1630, fresh from sorting out the heart, Harvey turned to what was then known as "generation," a broad heading that encompassed all the mysteries of sex, conception, and development. His study of life began with a plunge into death.

That was unavoidable, though it was ugly, because understanding the workings of the living machine required, as a first step, identifying its parts. This was the biological counterpart of disassembling a balky toaster or a misbehaving car engine and laying the parts out on a workbench. But this body shop was the real thing.

F ROM EARLY ON, HARVEY HAD HARBORED A PASSION FOR ANATomy and dissection. "The examination of the bodies of animals has always been my delight," he wrote, and he seemed untouched by the distaste for blood and gore that softer souls like Leonardo had to fight down. (Leonardo was so fervent an animal lover, according to his admirers, that he bought caged birds in order to set them free.) Any creature that hopped or flew or slithered or swam—some sixty species

---

*It takes about twenty seconds for blood to make a complete circuit of the body.

in all, by one biographer's tally—was likely to find itself pinned beneath Harvey's knife.*

Harvey probed the bodies of dogs, pigs, snakes, shrimp, frogs, oysters, and lobsters, as well as countless birds. He dissected hens by the flock, and studied their eggs with minute attention. Traveling through once-lush regions of France and Spain in 1631, a time when war and plague had emptied the countryside, Harvey lamented that "we could scarce see a dog, crow, kite, raven, or any other bird, or anything to anatomize." (As for the humans, Harvey noted grimly that "famine had made anatomies before I came.")

Harvey's fascination with dissection extended even to his own family. He observed the autopsy of one of his brothers and remarked with curiosity that he had "a spleen hanging like a letter V." A few months later, when Harvey's father died, Harvey dissected the old man himself. (This was one last good deed on the part of a proud parent, according to one biographer, "Thomas encouraging his son's medical studies in death as well as in life.") Peering intently at the organs glistening within his father's open body, William Harvey took special note of the "huge" colon.†

To crack the sex riddle, Harvey needed to find a way to frame the essential facts so boldly that the truth shone out. Science is the study of the real world, but the game always starts with turning away from reality for a time rather than confronting it. The trick is to find a stripped-down, simplified version of the world that can stand in for the real thing. Reality is too complicated to tackle head-on; a map is a better guide than the territory itself.

---

*Robert Boyle, one of the great figures of the scientific revolution, shared Harvey's obsession. A profoundly religious man, Boyle looked on dissection as a way of honoring God. Rummaging inside "dead and stinking carcasses" sounded unpleasant, he conceded, but in fact few pastimes were as "delightful" as exploring "the forsaken mansions of the omniscient Architect."

†As well as demonstrating Harvey's zeal, these particular dissections represented a practical, if unsentimental, response to the perennial problem of finding bodies to study.

Isaac Newton found that the cosmos could never be described tidily, for instance, as long as he pictured the Earth and moon as the immense, complicated structures that they are. But instead of thinking of them as huge, rocky lumps pulled this way and that by a myriad of other lumps, he imagined them as two isolated, featureless points in a mathematical drawing. Suddenly the heavens fell into place.

Harvey had mastered that game. Before he came along, the heart had lain wrapped in mystery. That darkness had stretched out for thousands of years. As soon as Harvey saw that "the heart is a pump," the mystery nearly solved itself. Now he meant to do something similar to explain sex.

As usual, the problem began with sorting out the role of women. What was it exactly that they contributed to the making of a new life? The trouble, which was a sign of a field in its infancy, was that Harvey had too many theories to pick from. He passed over the first, without even bothering to reject it explicitly. This was the age-old doctrine of woman-as-field, which implied that the woman made no contribution at all to the new generation. Harvey deemed that impossible, essentially on philosophical grounds, as a scientist today might reject the possibility that aliens built the pyramids. Whatever was true, *that* could not be.

That still left two theories of the case, and they didn't fit together. Both dated to ancient times. Both had hordes of clamorous supporters. Both rested on analogies that struck their partisans as self-evident.

One theory held that women were like men. Since men produced a fluid that played an essential role in conception, women must produce a similar fluid of their own. The second theory compared women not to men but to other female animals. Those animals produced eggs, the age-old symbol of fertility. Since women's reproductive anatomy resembled that of these egg-laying animals, women must produce eggs.

Harvey, a master of anatomy and an ingenious experimenter, belonged to the egg camp. Since ancient times, humans had watched chicks emerge from their eggs. Though no one had ever seen anything like an egg in mammals, many anatomists felt sure that such eggs must exist. So Harvey believed, fervently. At the time, his view was in the minority.

In the mid-1600s, Harvey's era, most medical authorities still echoed a view that Galen had set out fourteen centuries before. Women and men were variations on a theme. The two sexes were nearly identical in structure and function. They joined together in sex; they both carried on with enthusiasm and vigor; they both emitted fluids in the course of the action; they both reached orgasm. It stood to reason, Galen and his followers decreed, that if semen was the male contribution to conception, then the female contribution was some corresponding semen.

The one-sex model, as Galen's theory came to be known, held that women and men were anatomical twins. When it came to obvious features like eyes and ears and hands and feet, male and female plainly corresponded. The surprising claim of the one-sex model was that, structurally and anatomically, the sexual parts matched, too. This unlikely doctrine reigned for a millennium and a half.

The comparison required a bit of hard-to-follow scientific hand waving. With a bit of presto, change-o, testicles became ovaries, the penis became cervix and vagina, the scrotum became the uterus. Vesalius's *Fabric of the Human Body* is jammed with unsettling pictures meant to show, in the words of the historian Thomas Laqueur, that "the vagina really is a penis, and the uterus a scrotum."

As late as the 1700s, the most popular sex manual of the day held to the one-sex view, and even managed to put it in rhyme:

> For those that have the strictest searchers been
> Find women are but men turned outside in
> And men, if they but cast their eyes about
> May find they're women with their insides out.

The anatomists and physicians themselves had no doubt they had uncovered a vital truth, though perhaps one hard to convey. Laqueur compares their accounts to those you sometimes read today when scientists try to explain curved space and higher geometry, and go on about how a coffee cup and a doughnut are "really" the same because if they were made of modeling clay you could transform one into the other. For the public in centuries past, only halfway paying attention to such squabbles, the very obscurity of the purported explanations may have given them credibility, on the grounds that what sounds difficult must surely be deep.*

But the one-sex doctrine had more in its favor than an air of profundity. A man's testicles and a woman's ovaries did seem to match up, more or less—both came in pairs, both were near the belly, and both had something to do with sex. (For thousands of years, farmers and herders had castrated and spayed their animals. Mysteriously but reliably, this rendered a whole variety of creatures—cattle, horses, pigs, dogs, even camels—more manageable. More important, it kept them from reproducing without seeming to do any lasting harm.)

The evidence from spaying and gelding was not conclusive. No one could explain how it worked, for starters, and perhaps the procedures *did* do some subtle damage. You could imagine that surgery that injured an animal's heart, say, might make it placid and uninterested in mating. But that would not prove that the heart played a direct role in sex; perhaps the poor animal was simply too weak or wounded to carry on.

Still, even if the case for the one-sex model was not airtight, it seemed plausible. The rival camp spoke confidently of mammalian eggs, but, in these premicroscope days, no one could be certain such things even existed. Then what? If the "female testicles" didn't produce eggs, what could they be for, if not producing a female semen? This was seen as a strong argument, because everyone agreed that every structure in the body has a purpose. "Nothing is accidental in the works of nature," Aristotle had declared. "Everything is absolutely for the sake of something else." From the Greeks through Leonardo and Harvey, everyone echoed the point. Nature's designs, or God's, contained no unnecessary features.

---

*Centuries later, it would turn out that this outlandish doctrine had something to it: as a new embryo develops, both male and female genitalia *do* arise from the same tissue.

When it came to sex, this innocuous-sounding doctrine played out in surprising ways. Women's orgasms, for instance, could not just be a source of happiness; they had to have a purpose beyond mere pleasure. "When also in coition ye observe the same delight and concussion as in Males," asked the English scientist Nathaniel Highmore, in 1651, "why should we suppose Nature, beyond her custom, should abound in superfluities and useless parts?"

The unlikely consequence was that, in one remarkable way, conventional medical wisdom did right by women. This good deed was inadvertent, and it marked the very centuries when women were routinely maltreated or disdained. But if Galen was correct in saying that both women and men produced semen, then it followed that women, too, had to reach orgasm if they were to conceive. And the birth of children was hugely important. So women's orgasms were a matter of great concern, and not only to the women involved.

Across Europe and throughout the Arab world as well, physicians provided males quite specific advice on sexual technique. "Men should take their time playing with healthy women," wrote the Islamic physician Avicenna, in a celebrated work called the *Canon of Medicine*. "They should caress their breasts and pubis, and enfold their partners in their arms without really performing the act. And when their desire is fully roused, they should unite with the woman rubbing the area between the anus and the vulva, for this is the seat of pleasure. They should watch out for the moment when the woman clings more tightly, when her eyes start to go red, her breathing becomes more rapid, and she starts to stammer."

Sometimes ignorance *is* bliss.

PROPONENTS OF THE ONE-SEX MODEL BOLSTERED THEIR CASE with other observations. Even supposing for the sake of argument that women *did* have eggs, why were there two ovaries but, usually, only one baby? And how would those hypothetical eggs make their way to the womb? The Fallopian tubes seemed the natural route, but they did not quite reach all the way to the ovaries. Did the imaginary egg leap across that very real gap?

Wasn't it more likely that women produced a sexual fluid corresponding to a man's semen? Better yet, didn't the theory that two semens mix together explain why babies so often inherited features from both parents?

On the other hand, the argument in favor of the one-sex model had problems. Harvey ridiculed it. It was true, he agreed, that "during intercourse the male and female dissolve in one voluptuous sensation." But to say that both sexes melted deliciously was a long way from proving that both sexes produced semen. The theory could not be true in any event, Harvey snarled, because female genitalia were not on a par with the male's. "I, for my part, greatly wonder that from parts so imperfect and obscure, a fluid like the semen, so elaborate, concoct, and vivifying, can ever be produced."

That was name-calling, not science, but the one-sex model had confronted more serious challenges. It had even outlasted the "discovery" of the clitoris by an anatomist named Renaldus Columbus in 1559. The clitoris was "pre-eminently the seat of woman's delight," Columbus told his readers. "If it is permissible to give names to things discovered by me, it should be called the love or sweetness of Venus." This was a bold claim. Like Christopher Columbus before him, Renaldus Columbus had "discovered" territory that the natives had explored on their own eons before.

But, priority aside, this was confusing. Galen had declared that it was the vagina that was, anatomically, the counterpart of the penis; a second counterpart seemed too much of a good thing. Proponents of the one-sex model never flinched, though they did quarrel over just which bit of female anatomy was more akin to the penis.

They might have questioned their theory more vigorously had the whole debate not slipped away from specifics into a lament about the imperfect design of female bodies generally. Though the structures of the two sexes matched, Galen and his followers explained, the male versions were superior. The male was the template, the female a flawed and clumsy imitation, like a child's drawing of a Greek statue. (Through all these centuries, the female reproductive organs had no names of their own but were referred to by the names of their male counterparts.)

Physicians and biologists remarked again and again that the male apparatus was proudly exhibited while the female counterparts were relegated to the body's unseen interior, the fit home for the inferior and the undeveloped. Galen made a rueful comparison between a woman's reproductive organs, hidden away, and a mole's tiny, deep-set, and nearly useless eyes.[*]

Lest we reject Galen's views as the kind of ancient folly we have long since outgrown, we should note that nearly two thousand years after his death, physicians continued to draw damning conclusions from what they called female "interiority." "Their secret internal organs, women were told, determined their behavior," notes one history of nineteenth-century medicine. "Their concerns lay inevitably within the home."

THOSE THINKERS UNWILLING TO FULLY ENDORSE GALEN'S MODEL could opt instead for a modified version of his theory. This one had an even more impressive pedigree; it dated to Aristotle, a brilliant observer of the natural world and biology's founding father, who flourished around 350 BCE. William Harvey, disdainful of Galen but a devoted admirer of Aristotle, gave it a hard, serious look.

Like Harvey and Galen, Aristotle rejected the notion that women served merely as the field where the male seed grew. On the contrary, he insisted, women did contribute something physical and tangible to conception. But this substance was not a female semen. It was menstrual blood.

---

[*] By this reasoning, a man's flat chest would presumably have rated as inferior to a woman's protruding breasts, but that argument never made much headway. Aristotle had waved it aside impatiently, long before Galen. The superiority of the male's chest was self-evident, he argued, because it was firm and muscular, while the female's breasts were soft and spongy.

Galen had argued by analogy: male and female reproductive structures looked alike (to a generous eye) and therefore they functioned alike. Aristotle's argument relied on a different analogy. It, too, required a certain openness of mind. "Wait a second, hear me out," you can almost hear Aristotle plead, as his fellow investigators rolled their eyes and clutched their tunics in dismay.

For starters, semen and menstrual blood both seemed to play some part in the sex story, though it was by no means clear just what. (Detectives would have dubbed the human counterparts of such shady characters "persons of interest.") Aristotle rattled off points to ponder. Young boys did not produce semen, which appeared on the scene only when they reached physical and sexual maturity. Similarly, young girls did not menstruate, and neither did old women; menstruation overlapped precisely with a woman's child-bearing years.

Now Aristotle hammered home his argument. First, blood played a vital role in the body. Ordinary blood provided nutrition; menstrual blood must therefore fulfill some crucial mission of its own. Second, the embryo grew within the mother's body and was a tangible, physical structure that had not been present before the start of the pregnancy. Third, a pregnant woman no longer discharged the blood she had produced each month before her pregnancy.

The conclusion practically declared itself. Something appeared in pregnancy, and something disappeared. It should be plain to all, Aristotle announced, that the embryo was formed from the mother's menstrual blood. The semen shaped this raw material as a sculptor fashions clay. (As a bonus, the theory also explained the birth of females. Sometimes, Aristotle explained, "owing to youth, old age, or some similar cause," a man was too weak to produce proper, vital semen. If he didn't have enough oomph to beget sons, he would have to settle for daughters.)

FOR US, THIS FOCUS ON MENSTRUATION SEEMS PECULIAR. BUT OUR picture derives from the notion of menstruation as the flushing out of an unfertilized egg. The Greeks, who had never seen mamma-lian eggs, framed things quite differently.

Men and women had an inherent "vital heat," they believed, and it was this property of living organisms that set them apart from rocks and pots and other inert lumps. Though hard to define, heat was akin to vitality or soul or vigor. In modern terms, it corresponded more or less to metabolism. Aristotle, who favored homey images, talked of how the body nourished itself by means of a sort of "cooking."

Heat was a good thing, and men had more of it than women. "It rose naturally toward the heavens and towards the brain," writes the historian Merry Wiesner, "which explained why men, being hot and dry, were more rational and creative; women, being cold and wet, were more like the earth."* This was a far-reaching theory. It explained why women menstruated and men did not (men "burned up" their excess blood), and why men went bald and women did not (men "burned up" their hair), and why women had wide hips and men did not (women did not have enough heat to propel their flesh upward).

When it came to sex and reproduction, "cooking" transformed blood into semen by a process of purification and refinement. In wom-en's colder bodies, cooking never got properly underway and therefore achieved a less impressive transformation, merely converting ordinary blood into menstrual blood. Semen and menstrual blood were analo-gous, but semen was the high-grade product, menstrual blood its in-ferior counterpart. Semen was rare and valuable, a nectar made in tiny quantities. Menstrual blood was raw and unfinished, a crude and overabundant home brew. "In a weaker organism," Aristotle explained, "there will inevitably be a greater flow of less fully cooked blood."

This half-baked theory won wide support from the era of the Greeks to the time of Harvey and the king's deer. Aristotle believed he had dealt a death blow to the idea of female semen, which had floated around long before Galen spelled out the details. Since women were

---

*In the Middle Ages, this belief in women's "earthy" nature transformed into the view that women were sexually voracious. More lustful than men but less rational, women had to be held in check by men, for the sake of order and propriety.

cold, it made sense that they produced menstrual blood, which was raw, rather than semen, which was cooked. How could anyone believe that women produced *both*?

To imagine such a thing, Aristotle scoffed, would be to claim that "woman is at one and the same time of hot and cold temperament, which is the height of absurdity." One might as well claim that a stone could sit atop the water, because it was so light, and simultaneously sink beneath the waves, because it was so heavy.

The way that conception worked, as Aristotle set it out, was that both semen and menstrual blood played vital roles, but it was semen alone that shaped the new life. Every handmade item you might name—a loaf of bread, a clay jug, a stone house, a wooden chair—involved a craftsman transforming raw material. For living creatures, Aristotle proclaimed, exactly the same pattern held: a creative, shaping force took a bit of humdrum stuff and transformed it.

One sex performed magic; the other provided supplies. Three guesses.

PART TWO

THE SEARCH FOR THE EGG

*"How often have I said to you that when you have
eliminated the impossible, whatever remains,
however improbable, must be the truth?"*

—Arthur Conan Doyle, *The Sign of the Four*

SEVEN

MISSING: ONE UNIVERSE (REWARD TO FINDER)

B OTH OF THE MAIN ATTEMPTS AT EXPLAINING THE MYSTERY OF
sex and babies—Galen's one-sex, two-semens model and Aristotle's semen-and-menstrual-blood variation—had endured from before
the birth of Christ to the dawn of the modern age. Both seemed a
mix of the plausible and the bizarre (*men and women the same? embryos built of menstrual blood?*). Neither side claimed its case was airtight. Aristotle, who was a brilliant and tireless student of the animal
kingdom, knew perfectly well that many female animals did not menstruate. How could it be, then, that their embryos were formed of
menstrual blood? The answer he came up with was that these animals
did produce various secretions, and those fluids played a role akin to
menstrual blood.

The argument smacks of special pleading. But even if we put animals aside and look only at humans, comparing semen and menstrual
blood was a stretch. What was the similarity, really, between a once-
a-month fluid and one that appeared each time sex took place? As we
have seen, the ground that Galen stood on was just as shaky. What *was*
the true resemblance between a uterus and a scrotum?

It was impatient, dyspeptic William Harvey who proposed to break the impasse. He would do it not by talking but by experimenting. Harvey had a flair for designing experiments to resolve questions that had seemed doomed never to move beyond insult and guesswork. Early in his study of the heart, for instance, he had been stymied by just how hard it was to see what was happening. He had looked inside the chest of a living dog and seen its heart contract and expand (any dog lover would grow queasy at the details of these hideous experiments), but the sheer speed of the action made it impossible to understand the big picture.

Then it had dawned on Harvey to look at cold-blooded creatures like frogs, snakes, snails, and shrimp. He found with a thrill—the discovery presumably set *his* heart racing—that their hearts beat ever so slowly. He dissected countless amphibians and reptiles. Centuries before the invention of the slow-motion camera, Harvey had devised a way to slow down the beating heart so that he could study it in detail.

This strategy of approaching a problem by way of an easier example would become one of Harvey's favorites. It had brought him ridicule—*What do lowly snakes have to do with noble humans?*—but in the end it provided insights into the heart that had eluded all his predecessors. Now, with the mysteries of reproduction on his mind, Harvey took advantage of his friendship with the king. He set out to explore the royal menageries.

The ostrich, the world's largest bird, caught his eye at once. The bird's style of mating was as striking as its size. Harvey looked on, agog. "I myself have seen a hen ostrich, when her keeper gently stroked her back with the intent to arouse her desire, throw herself on the ground . . . and disclose and stretch out her vulva," he wrote. "When the cock bird saw this, being instantly enflamed with desire, he mounted her, and with one foot on the ground and the other pressing on her back as she lay, accomplished his purpose with an exceedingly large and vibrant yard [i.e., penis] that you might have taken for a [cow's] tongue. All this went on with much muttering and noise on both their parts, stretching out and pulling back their heads and many other signs of rejoicing."

When captive ostriches happened to die, Harvey hurried to dissect them, apparently on the theory that the huge birds would serve as a kind of large-print primer on the secrets of reproductive anatomy. The appeal of this strategy was hard to deny—an adult ostrich and an adult human are roughly the same size, but an ostrich egg is larger than any other animal's, about five inches in diameter. (The eggs of all mammals, from dachshunds to giraffes to humans, are virtually the same size.) Perhaps what was hidden in the human being would proclaim itself in the giant bird.

But birds turned out to be an uncertain guide to human anatomy. Though male swans, ducks, and ostriches have penises, for instance, the males in many smaller species do not.* At the end of all his studies, Harvey found himself well-informed on bird anatomy but unsure what to make of his newfound knowledge. "In a black drake I once saw a penis of such length that, after coition, a hen pursued it as it trailed along the ground and eagerly pecked at it, believing it, I am sure, to have been a worm, and this made the drake retract it more quickly than is his custom."

Earlier observers had found themselves not merely confused by the complexities of sex in birds but positively outraged. In 1474 a huge crowd had gathered in Basel, Switzerland, to see a rooster burned at the stake. The bird's sin was that, even though he was to all appearances a bold, strutting, cock-a-doodling male, "he" had laid eggs. This was contrary to nature and dangerous besides—the eggs of such hybrid creatures were well-known to hatch into half-bird, half-serpent creatures whose mere glance could kill—and execution was the only fitting response.

Oddly, Harvey had made a similar mistake about the sex of a bird. Harvey's wife had a beloved pet parrot who sang on command and sat in her lap to have his head scratched. (This meager fact is virtually all we know of Elizabeth Harvey. It is tempting to imagine her patiently

---

*Nature is endlessly innovative, and exuberantly so when it comes to sexual anatomy. The octopus has no penis at all (though it does have three hearts), for instance; snakes have two penises, which they use one at a time, alternately in successive matings; and several kinds of marine flatworms have dozens.

reciting, "Who's a pretty bird?" while her husband sliced up cats and dogs and mice in the next room.) After many years, Harvey recalled, the bird "fell sick, and by being seized with repeated attacks of convulsions, died, to our great sorrow, in its mistress's lap, where it had so often loved to lie." Harvey knew that only male parrots speak and sing. He dissected his pet to see what it had died of and found, to his amazement, "an almost complete egg in its oviduct."

Armed with countless but confusing bits of information from the royal menageries, Harvey needed to find a better way to demonstrate to the world that mammals did have eggs, as he had long believed. This was the point of his deer-hunting excursions with the king.

The king's deer, which mated and conceived on the same predictable schedule year after year, seemed ideal research subjects. A devoted admirer of Aristotle, Harvey felt certain that he would confirm his master's teachings: when he examined the bodies of newly pregnant deer he would find within the uterus a small, egg-shaped embryo fashioned from the female's menstrual blood by the male's semen.[*]

But he didn't. He found nothing at all. He looked again. Still, he found nothing. He looked at pregnant dogs and found nothing. At rabbits, nothing. Stymied in his search for semen, menstrual blood, or embryo—for any clue he could touch—Harvey drew the only possible conclusion. Females conceived "as by Contagion."

In a world that had seen whole countries devastated by plague, this was perhaps a natural thought. Harvey was not invoking magic. He pointed out that "epidemic, contagious, and pestilential diseases scatter their seeds and are propagated to a distance through the air." If plague could scythe through whole cities, passing from one person to another invisibly and unaccountably—without a sting or bite or wound—why could not semen do its fertilizing work at a distance, without physical contact?

Semen evidently exerted its influence from afar, as a conductor's baton guides an orchestra. Its physical absence in the female's body

---

[*]Naturalists would later learn that deer do not menstruate. Among nonhuman females, menstruation occurs almost solely in primates (and bats).

testified to its power. This potent fluid contained something, Harvey wrote, "analogous to the essence of the stars."

Harvey cited other analogies. Magnets were much in vogue in the 1600s, because they produced a force that was both undeniable and inexplicable. Magnets could draw iron filings across empty space; why should semen, too, not act at a distance?

Most significantly, perhaps, Harvey compared the uterus and the brain. The brain secreted thoughts; an idea was often described as a "conception" of the brain. It was telling, Harvey claimed, that we used the same word when we talked about the "conception" of new life. An embryo was a "conception" of the uterus. Neither "conception" was a physical object. No wonder, then, that no matter how closely you examined a newly opened uterus, you saw nothing.

T EXTBOOK ACCOUNTS OF SCIENCE SAY THAT YOU TEST A THEORY by performing an experiment. If the result contradicts the theory, you throw the theory away and think up a new one. But it seldom works that way. Harvey sliced open his deer in the full expectation that he would see a tiny embryo that he could flick into the sunlight with the tip of his knife. When he didn't find it, he scratched his head in bewilderment, but he stuck to his theory. It *had* to be right. Sooner or later, he knew, he would see how.

Modern scientists display the same reluctance to abandon trusted theories, in even the most daunting circumstances. When the *New York Times* reported that "Ninety Percent of the Universe Found 'Missing' by Astronomer," the story made page 1, but scientists blithely carried on with business as usual. The *Times* noted merely that the great figure in the field, a Princeton physicist named Martin Schwarzschild, "hopes the missing matter in the universe can be located."

When Harvey peered into his deer, perhaps the prudent conclusion would have been, "I cannot see anything and therefore I cannot conclude anything." But one of Harvey's greatest coups had come from taking precisely the opposite line. In tracing blood's voyage around the body, he had lost sight of his quarry: he knew that the heart pumps blood into the arteries, which nourish the body's tissues, and he knew that eventually blood returns to the veins, which carry it back to the heart. But, limited by the technology of his era, he could not see how blood made its way from the arteries to the veins.

He deduced—correctly and brilliantly—the existence of a network of blood vessels too tiny to make out with the naked eye. They are now called capillaries, and they were first seen only after Harvey's death. "I cannot see anything," he had reasoned, "and I know exactly what to conclude from that."

When it came to sex and conception, too, Harvey carried on undaunted by his quarry's vanishing. (We now understand how Harvey's deer experiments misled him. Harvey could never have known it, but deer might as well have been designed to hoodwink a scientist. Deer embryos don't have the familiar round shape that Harvey expected from dissecting rabbits and dogs; they are long, slender, strand-like, and easy to miss even when you know what to look for.

Worse, there is typically only one embryo per doe, and that one hardly begins to grow until ten days after conception. Harvey did notice "mucous filaments like spiders' threads," but he did not give them any special notice. If the king's hunting parties had been designed to pursue rabbits rather than deer [and if Harvey had waited a bit before cutting open the pregnant rabbits], the trophies in the royal lodge would have had an odd look, but Harvey might have found what he longed to see.)

In his old age, Harvey published his second great work, *Disputations Touching the Generation of Animals*. He was seventy-three, famous but in bad health and in bad spirits. Tortured by gout, he tried to ease the pain by submerging his legs in tubs of ice water "till he was almost dead with cold," in the words of an acquaintance. Then he would stumble to the stove as fast as he could, to thaw out.

These were harsh times. With his beloved king executed and many of his own research papers destroyed during the civil war, Harvey had opted to retreat from the world. (By one account he tried to kill himself by swallowing opium.) A friend persuaded him that his decades of work should not go unrevealed, and Harvey handed over the manuscript of the *Generation of Animals*.

This took considerable coaxing, for Harvey still grew indignant when he recalled the scorn that had greeted his book on the heart. That criticism had long since turned to adulation, but Harvey had not mellowed. Most infuriating of all, his admirers had misinterpreted the message of *On Circulation* and placed Harvey at the head of an army he despised.

The world hailed him as the man who had proved that the body was a machine. This was nonsense, Harvey thundered. He wanted nothing to do with those who would banish God and spirit from the world in favor of soulless pipes and valves. Those who would reduce the living world to mere mechanism were "shit-breeches."

Harvey had ventured out ahead of the evidence when he proposed that the body has capillaries, and in the new book he did so again. His claim, destined to become famous, was that all life comes from an egg. This was a sweeping assertion, embracing egg-laying animals like birds and live-bearing mammals like dogs, cats, and humans. "All animals whatsoever . . . nay, man himself, are all engendered from an egg," Harvey wrote.

In a poem published as the dedication to the book, an admirer of Harvey's put the claim even more boldly. In the ancient battle between analogies—*Were women like men, or were women like other female animals?*—the poem left no doubt about Harvey's view:

> Both the hen and housewife are so matched,
> That her son born, is only her son hatched.

A drawing on the title page depicted Zeus on his throne, holding open an immense egg, one half in each hand. A horde of living creatures spilled out: a bird, a baby, a spider, a butterfly, a fish, even (despite its obstinately missing embryo) one of the king's deer. Zeus's egg bore a Latin motto, forever after associated with Harvey. *Ex ovo omnia*, it read. *Everything comes from the egg.*

The motto makes Harvey sound more modern than he was. He didn't mean that the egg was the female's contribution to the new life she helped create, the counterpart of the male's semen. The heart of his claim was that humans and all other mammals begin their lives as fragile clumps of tissue, as tiny, naked eggs that grow within the mother's body. (Harvey and his contemporaries jumbled together different senses of the word "egg" in a way that confused them and still confuses modern readers. When we talk about a woman's eggs today, we have in mind the sex cells that unite with sperm cells to form embryos. Harvey did not know about sex cells. For him, "egg" meant "tiny embryo." In addition to that conceptual and linguistic muddle, Harvey had made a significant anatomical mistake: he discounted the role of the ovaries and believed that eggs originated in the uterus.)

Just what the woman's contribution to pregnancy was, Harvey admitted, he had not managed to learn. And his analogies—conception was like infection or magnetism or thought—did "but replace one mystery by another." As frank as ever, he spelled out his failure in plain language. "There is no sensible [i.e., detectable] thing to be found in the uterus, after coition," he wrote in exasperation, "and yet there is a necessity that something should be there." He bristled for a moment—no one else had come close to finding an answer "even in his dreams"—but then he conceded that he had not, either. He had no choice "but to confess myself at a standstill."

But in several ways Harvey *had* moved the case decisively along. So widely admired was he by this time, for starters, that he had shifted the debate. By focusing so intently on the egg—even though his notion of the egg was hazy—he had in effect directed the next generation of scientists to find those eggs at all costs.

Next, Harvey had ended the reign of his hero, Aristotle. Whatever would eventually turn out to be true, the semen-and-menstrual-blood theory was false. In the detectives' case room, the picture of Aristotle now bore a large X.

Harvey's last lesson was the most important. The way to take on this mystery was to stop talking and start experimenting.

EIGHT

SHARKS' TEETH
AND COWS' EGGS

O N July 15, 1669, just a decade after Harvey's death, an unlikely package arrived at the headquarters of the Royal Society, in London. Inside a small glass jar was a tangled snarl of skinny, floppy tubes floating in a clear liquid, like a miniature pasta dish gone awry. A proud note resolved the mystery of the jar's contents. The Royal Society had before it, according to one Regnier de Graaf, "the testicle of a dormouse, unraveled by my method."

This dormouse preparation sounds like a spoof of an unlikely venture, as if someone had gathered shavings from a unicorn's horn. In truth it resolved a basic question in anatomy, and it showed how skilled de Graaf was in sorting out such riddles. A young, self-assured Dutchman—he was only twenty-eight when he sent his gift to the Royal Society—de Graaf was a physician with a special interest in anatomy and the mysteries of reproduction. Born when Harvey was already an old man, de Graaf and a handful of rivals would claim the next era in the sex and reproduction story for themselves.

Harvey had guessed right about the crucial importance of eggs, but he had not proved his case. That giant task still lay ahead. Still unresolved, too, were a host of fundamental questions concerning sexual anatomy. Taking on those mysteries would be the mission of de Graaf and his frenemies. They would grab the baton from Harvey and race ahead, happily opening up new territory at one moment, knocking off old riddles at the next, and mocking their foolish peers in between times.

A handsome, square-jawed, outgoing man, de Graaf subverted the cliché of the shy and awkward scholar. Despite a starry record in his university days, he never held an academic post, presumably because of the prejudice he faced as a Catholic in Protestant Holland. Instead, he practiced medicine in Delft. His practice thrived, but he truly came alive only when he had sent off his last patient of the day. Then he turned to his own research.

Until de Graaf took up the dormouse, no one had truly under-stood the structure of testicles. Were they some sort of solid, bean-like gland? The answer, de Graaf demonstrated, was that the testicles were made up of countless, tangled tubules. Take those tubules from a dog, say, and try to disentangle them, and the result was a hideous mess.

But try the same experiment with a dormouse,* and "you will behold a delightful and surprising sight." Put the tubules in a pan of water, and they separated and unraveled by themselves. "I often demonstrated this to the physicians and surgeons of this city," de Graaf boasted. His audiences gossiped so excitedly about what he had showed them, he went on, that "I became afraid lest, because of the laziness of the engraver, another should snatch from me the glory of this magnificent discovery" before he'd had a chance to publish it himself.

De Graaf set out his anatomical discoveries in two volumes, one on males and one on females. Both were clear and down-to-earth, espe-cially in contrast with Harvey's sometimes obscure prose. "Whatever is procreated of the semen properly so called originates and is perfected either in the same place or in different places," Harvey wrote, in one

---

*De Graaf studied the edible dormouse, which resembles a squirrel. (It is larger than the dor-mouse that played a cameo role at the Mad Hatter's tea party in *Alice's Adventures in Wonder-land*.) Its name reflects the Roman practice of roasting dormice and dipping them in honey.

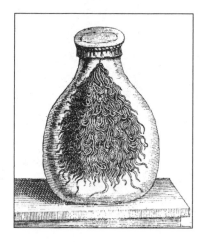

FIGURE 8.1. De Graaf's "magnificent discovery": the structure of a dormouse testicle revealed.

typically dense passage. De Graaf was not as arcane. The male genitals consist of "the part that enters the vagina, and the one which remains outside hanging down like a bag."*

That plainspoken tone was characteristic. "The pleasure of copulation is indescribable," de Graaf remarked, and taking his own advice, he moved on at once to other matters. He was similarly brisk when considering time honored but "absurd" doctrines, like Hippocrates's claim that sperm from the right testicle produced males and, from the left, females. De Graaf told of "a citizen of Delft who possessed only a right testicle and yet was presented with various daughters by a most honorable wife."

He had just as little time for Aristotle's claim that the testicles served only as counterweights to keep the sperm ducts untangled. De Graaf had seen illnesses and injuries to the testicles that rendered men sterile; those men still had counterweights, but they could not father children. And what of birds and other animals whose testicles

---

*The human penis, though of compelling interest to its owner, is of far plainer design than many others in the animal kingdom. While a barnacle stays glued to its rock, for instance, its bristle-covered, chemical-sensing penis can stretch this way and that in search of a mate in the neighborhood. The barnacle penis extends to eight or nine times the animal's length. It is the biggest in nature, proportionally, and moved Darwin to wonder. The human penis is essentially a tube to deliver semen. (Its role in excreting urine is apparently an evolutionary afterthought, an improvised two-fer in the cobbled-together fashion that turns up so often in nature.)

were inside their abdomen? In that position, they "could not possibly act as weights," and yet the animals reproduced perfectly well.

Combative and sarcastic, de Graaf flailed energetically at his enemies. "I am certainly surprised that men of outstanding ability can still be mistaken in such an obvious matter," or, more bluntly, "Your book came out of your arse, not your head." At times he snarled in exasperation. "Certain people want to know how semen, which has its origin from such very red blood, grows milky white." *Who had time for such questions?* "This should no more cause surprise than the fact that milk is white, although it draws its origin from green plants."

Even worse than critics were poachers who ventured too near de Graaf's own discoveries. He had sent his dormouse to the Royal Society in the first place because he had heard the "astonishing" news that English scientists claimed they had found the structure of the testicles before he had. They had done no such thing. Their accounts of their supposed discoveries had to be "twisted like a waxen nose" before anyone could find any sense at all in what they had said. *Compare their vague words with the magnificent contents of this bell jar!*

T HE REASON FOR THE RANCOR WAS COMPETITION. THE SEX MYStery was in the air, and *someone* was going to make his name. De Graaf had many rivals; two of the most accomplished were slightly older contemporaries he had come to know at Leiden University in Holland. Both were destined to make major contributions to science, to abandon science at the peak of their careers so that they could devote themselves fully to God, and to die young.

Nicolaus Steno was Danish, a physician by training but a man of such varied and shifting interests that he pleaded with God for steadfastness. "I pray thee, O God, take this plague from me and free my soul of all distraction, to work on one thing alone," he wrote in his journal. God paid no heed. Medicine's main rival for Steno's attention was geology. He was the first to explain why fossilized seashells turn up on mountaintops, and he showed that the Earth is ancient (though he never said outright that the customary Bible-derived age of a mere six thousand years could not be correct). He pursued medicine and geology almost simultaneously, dashing between investigations. His skill in dissection was so formidable that one dazzled spectator, who had watched him tease apart the eye of a horse, exclaimed, "He would count the bones of a flea—if fleas have bones."

In 1666 Steno's various interests converged. A fishing boat off the Italian coast hauled in a great white shark. Steno, who was twenty-eight at the time, was working at the Medici palace in Florence, where the grand duke supported a kind of informal scientific academy. So fascinating a trophy surely had to be investigated in Florence. There Steno dissected the head of the 2,600-pound monster.

He focused primarily on the great beast's teeth. They bore a striking resemblance to well-known, reputedly magical stones called "tongue stones." Folk wisdom had it that the stones fell from the sky on moonless nights (which explained why no one had ever seen it happen). Steno had no time for old myths. He held a shark tooth and a tongue stone next to one another and showed that they resembled one another as closely "as one egg resembles another." Tongue stones, Steno explained, were fossilized shark teeth.

More important, Steno explained how it could be that a farmer plowing a field miles from the ocean might unearth signs of life from an ancient sea. The earth was restless, Steno wrote. Wait long enough, and seas might rise up and mountains tumble down.

Steno wrote up his findings for the grand duke and included them as a sort of afterword to a longer work on the anatomy of muscles. Almost as an afterthought, he added still another few pages. This section, too, veered away from what had come before, though Steno did discuss sharks once more. This time he focused on reproductive anatomy, based on his dissections of a different species of shark, the dogfish, and also of stingrays.

Sharks give birth to live young, and rays lay eggs. Even so, Steno found himself struck by the similarity of their reproductive tracts. Then he recalled the appearance of the corresponding structures in women.

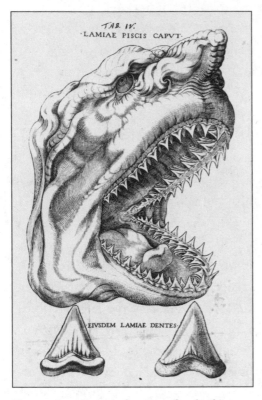

FIGURE 8.2. Steno's drawing of a shark's jaw, and a tooth seen from the front and back.

"Having seen that the testicles of viviparous [live-bearing] animals contain eggs, and having noticed that their uterus opened into the abdomen like an oviduct," Steno wrote, "I have no doubt that the testicles of women are analogous to the ovaries." This was a brilliant and daring leap, but, as Steno noted, it relied entirely on analogy. Humans and other animals had structures that *looked* alike, and therefore they *were* alike.

Steno had "no doubt" that women have eggs, but neither he nor anyone else had ever seen them. Nor had he made any attempt to solve the mystery of how the eggs made their way from ovary to uterus. But he had made a claim, in print, before anyone else. Steno had planted a flag.

Jan Swammerdam was another brilliant anatomist, a friend of Steno's almost exactly the same age, and perhaps even more pious and more tormented by guilt. A Dutchman like de Graaf, Swammerdam had a full measure of his countryman's ambition and competitiveness but none of his rowdiness.

Swammerdam had a newly minted medical degree, like de Graaf and Steno, but he never practiced. Instead, he pursued his own anatomical investigations with the obsessiveness that marked all his endeavors, while relying on his father's grudging financial support. In his marathons at the dissecting table, Swammerdam neglected his prayers. Recalling his misdeeds, he burst into tears. "For it was as if a warring host were there within my spirit, the one party compelling me to cling to God, the other, with infinite arguments, to go on in my [pursuit of] curiosities."

In 1667, at the same time that Steno was pondering shark teeth and human ovaries, an anatomy professor from Swammerdam's university days recruited him to help with a dissection. The two men set to work dissecting a female corpse, paying special attention to the uterus and Fallopian tubes. Like Steno, they made a point of noting the strong resemblance between the ovaries in humans and in egg-laying animals. Like de Graaf (although a few years before him), they observed and dissected small protuberances within the ovaries that they decided were the long-sought eggs (or perhaps structures that contained the eggs).

Swammerdam, whose drawing skills matched his formidable anatomical gifts, prepared illustrations. He went further, using a technique he had mastered for preserving organs in alcohol and injecting their vessels with red or yellow wax, which produced lines like the roads on a map. One lovingly preserved uterus made its way to the Royal Society, where it occupied a place of honor.

Almost inevitably, the friendship that had once bound de Graaf, Steno, and Swammerdam turned to rancor and jealousy as they squabbled over priority. They scrambled to get something into print and flung insults and accusations back and forth. In the long view of posterity, the stakes seem lower. All three men had grasped the same great truth—women have eggs—at virtually the same time. But it was Regnier de Graaf who not only saw the truth but devised an experiment to show it.

NINE

THE EGG, AT LAST

P ROUD AS HE WAS OF THE DISCOVERIES HE HAD REPORTED IN HIS book on male anatomy, De Graaf knew that it would be the companion volume, on females, that would make his name. *A New Treatise Concerning the Generative Organs of Women* appeared in 1672, four years after his volume on men. De Graaf dedicated the new work to Cosimo III, the Grand Duke of Tuscany. Such wooing of the wealthy and powerful was customary. But de Graaf also meant to claim his place in a daunting lineup.

Galileo had dedicated one of *his* great works, *The Starry Messenger*, to Cosimo II, also a Grand Duke of Tuscany and the grandfather of Cosimo III. *The Starry Messenger* contained the first report on a new invention, the telescope, and announced a startling picture of the heavens. Unlike Galileo, de Graaf wrote in his dedication to the grand duke, he had not gazed on distant stars. But he, also, had done great things. He had "made bold to strip off Nature's robe"—it is hard not to picture Mother Nature crouched over and trying to cover her naked body with her hands—and thereby revealed "the whole workshop of human manufacture and its tools."

The new book was as ambitious as its dedication implied. First, de Graaf took an axe to conventional wisdom. He scoffed at the one-sex, two-semen model: "The notion of some people that the vagina corresponds with the penis of males, differing only in being inside rather than out, we say is ridiculous. The vagina bears no similarity at all to the penis."

Nor were the fluids produced by the two sexes comparable. De Graaf noted, with his customary briskness, that "certain females . . . with lascivious thoughts, frisky fingers, or instruments devised contrary to decent morals, wickedly stir themselves up to such a pitch that they eject copious quantities of this kind of matter." ("In libidinous women," he added, "the mere sight of a handsome man" was enough.) But this fluid was not semen, de Graaf insisted, and he moved on.

He began with a no-frills summary of sexual ABCs. First came a remarkable tribute to God the designer. "The woman's vagina is so cleverly constructed that it will accommodate itself to each and every penis, it will go out to meet a short one, retire before a long one, dilate for a fat one, and constrict for a thin one. Nature has taken account of every variety of penis, and so there is no need solicitously to seek a scabbard the same size as your knife. Through the beneficence of the Creator you can find one anywhere."

He debunked folk beliefs. Menstrual blood did not turn crops sterile or do any other harm; "menstrual blood is of its nature a benign fluid." The size of the nose had nothing to do with the size of the penis. Modern anatomists like Columbus and Falloppius, who both "claimed the glory" of having discovered the clitoris, had done no such thing. The clitoris was old news, de Graaf declared, though it had been mislaid for two thousand years. Hippocrates and other ancient writers had known all about it, as had women since the dawn of time.

De Graaf proceeded to consider the organs directly involved in conception and pregnancy. His discussion of the uterus almost exactly anticipated that of the twentieth-century anatomist Frank Gonzalez-Crussi, who lamented that humankind's first home has "a most unfortunate topography: a bag of urine in front, a repository of excrement behind." De Graaf's take was more philosophical. The

uterus was "set between the bladder and the rectum as though between two pillows," he noted. "Some have thought that the purpose
was that men, who are born of fragile and perishable material between dung and urine and are destined to be resolved into earth
and cinders, should, when they remembered their vile and abject
condition, fold the wings of their pride."

The uterus itself contained mysteries that would further dampen
the pride of any would-be investigator. How did this organ, shaped
"like a pear which has been slightly squashed," perform its nurturing
role? How did it know when the time had come to deliver the baby
within it? "Writers of the sharpest wits . . . cannot understand it," de
Graaf conceded. "Thus, that which comprehended these writers does
not allow them to comprehend it."

All this brought de Graaf to his most important point and moved
him a crucial step ahead of the formidable William Harvey. What Harvey believed—that mammals had eggs—de Graaf had *seen*. He had
looked at all sorts of animals, egg layers and live-bearers, and all of
them "have ovaries full of eggs, and the eggs of the [mammals] are
fertilized and reach the uterus in the same way as in birds." (And de
Graaf's eggs, it is worth noting, were eggs in the proper, biological, sex-
cell sense of the word.)

De Graaf made sure no one could miss his point: what was true of
other female animals was true of women, as well. "Since all this can be
observed in cows, ewes, and other animals which I have dissected in
large numbers, everyone will admit that in women, who also have eggs
in their 'testicles' and the tubes annexed to their uterus, as the brute
beasts do, generation takes place in the same way."

These were daring arguments, even with Harvey as a shield. De
Graaf knew that his critics were poised to burst into laughter at the
mere mention of women and eggs, but he warned that they would look
"frivolous and stupid" if they did not heed him.

He turned his wrath toward those writers who viewed women as
incomplete, inferior men or as ornaments whose role was to beautify
the world "like the peacock's tail." This was "ridiculous" and insulting

both to women and to God. "Nature had her mind on the job when generating the female as well as when generating the male," de Graaf thundered.

Such fervor in defense of women was unusual.* Typically, in the early modern age, women were condemned for their lustful, fickle natures. Men were supposedly higher-minded. Certainly this was the view of Robert Boyle, one of the founders of the Royal Society and the most important English scientist in the generation before Isaac Newton. Celibate through his long life, Boyle feared the wanton, scheming ways of women, all of them temptresses like Eve: "I am confident that thousands would be whores could they but be so without being thought so." Men's highest calling was to study God's works, but women would lure the weak and unwary from that sacred mission. Who would peer through a microscope, Boyle asked, when he might be staring down a lady's cleavage?

De Graaf had little use for such attacks. Focused on his work, he called on anatomists to amend their language, as a first step: "The common function of the female 'testicles' is to generate the eggs, foster them, and bring them to maturity. Thus, in women, they perform the same task as do the ovaries of birds. Hence they should be called women's 'ovaries' rather than 'testicles,' especially as they bear no similarity either in shape or content to the male testicles."

De Graaf paused for breath, then sped across the finish line: "On account of this lack of similarity, they have been regarded by many as bodies without function; quite wrongly, because they are absolutely essential for generation."

D E GRAAF HAD SEEN THE BIG PICTURE: WITHIN THE OVARIES were eggs, and those eggs, in some mysterious conjunction with

*Bias against females turned up in the least likely settings. Observers from Aristotle to Leonardo—this was a span of eighteen centuries—claimed, for instance, that they had noticed a striking pattern. Male chicks came from round eggs, which were close to the perfect shape; female chicks came from eggs that were long and pointy and, therefore, inferior.

semen, formed embryos. And yet de Graaf had *not* seen eggs. Not quite. He believed sincerely that he had. What he had actually seen were the follicles that contain the eggs. Today those small, bumpy structures embedded inside the ovary are called Graafian follicles. A follicle ruptures and releases the egg within, which travels through the Fallopian tubes. (Vesalius, anatomy's founding father, had observed these follicles a hundred years earlier but had dismissed them as signs of infection rather than anything to do with conception. We will run into many similar examples in the course of our story, where a brilliant investigator notices a smoking gun, picks it up, mulls it over, and then tosses it aside as irrelevant.)

With a series of elegant and bold experiments, de Graaf came within inches of piecing together the whole egg story. He began with experiments akin to Harvey's dissections of deer, though he used rabbits instead. But unlike Harvey, who had failed to see any changes in the ovaries, and therefore discounted them, de Graaf saw changes galore.

In the first few days after mating, the follicles in the ovary had reddened and swelled and then burst. Staring intently, de Graaf grew to suspect that something within the follicles "had been disrupted or expelled." (In theory, de Graaf *might* have managed to spot an egg with his naked eye, but it would take his successors, armed with microscopes, another 150 years to do so. We think of eggs as easy to find, but that is because we picture bird's eggs. Those eggs are enormous, since they have to provide all the nourishment that the developing chick will ever get.)

Though he had missed the release of the egg itself, de Graaf continued his surveillance. Now he kept careful watch on the uterus, too. His tools were little more than sharp eyes and the ability to count. Instead of dissecting his rabbits only a few hours or even a few days after they had mated, he waited several days. Now he found ruptured follicles in the ovary *and* tiny embryos in the uterus. De Graaf asked the crucial question: *How many of each?*

To his delight, the number of ruptured follicles matched the number of embryos! Detectives on the track of a killer could hardly have put together a more promising case. It was as if the police had been trailing a murderer who liked to poison his victims. On a night when their suspect had been caught with two empty vials in his overcoat pocket, detectives had found two victims. On another night, three vials, three victims.

Then came trouble. De Graaf cut open a rabbit that had mated six days before. He counted ten ruptured follicles but only six embryos! This was not a problem, de Graaf insisted. The key was that he never saw *more* embryos than ruptured follicles. Most likely, de Graaf proposed, things simply went wrong sometimes—by random misfortune, the missing embryos had "come to a sinister end." That argument carried weight. Everyone knew that human pregnancies often end in miscarriage. Crisis averted.

De Graaf had moved well beyond Harvey. His biggest advance was conceptual. When Harvey talked about an egg, he had in mind a living organism in its earliest stages and *not* the female's contribution to a new, living creature. That confusion of "embryo" and "egg" had muddied the picture. De Graaf cleared things up and convinced the world to shift focus. In his picture, the egg emerged from the female's ovary and combined somehow with semen from the male to form a new organism.

This was an especially bold claim, since de Graaf had never quite managed to see the egg directly. He came to his insight partly by virtue of intellectual daring and partly by good fortune. Harvey's choice of deer made trouble for him, as we have seen. De Graaf's choice of rabbits made matters clear, and at times misleadingly clear. It so happens that in rabbits the act of mating stimulates the female to release eggs.[*] Watching his rabbits, de Graaf witnessed a straightforward series of events: a female mated, follicles within her ovary changed, the follicles ruptured, embryos developed, and the number of embryos matched (usually) the number of ruptured follicles.

But for most mammals, mating does *not* induce ovulation. If de Graaf had happened to choose a different animal to study—if, for

---

[*] A female rabbit can give birth to a litter and be pregnant again within twenty-four hours. Hence, "breeding like rabbits."

instance, he had somehow managed to look at human beings—he would have seen that whether an egg is released has nothing to do with mating. The pattern with women is not so tidy as with rabbits—virgins release one egg a month, and so do women with sexual partners—and de Graaf might have joined the throng of befuddled scientists unsure just what eggs had to do with conception.

As it was, he made errors aplenty. The biggest was following Harvey when it came to thinking about semen's role. For both these scientists, semen retained its mystery because they could not find it anywhere within the bodies of the female animals they dissected, even knowing it had to be there. Instead, they talked of "seminal vapor" and "irradiation," and pictured semen somehow manipulating the process of conception like a magician waving his fingers and making a silken handkerchief across the room dance and flutter.

D E GRAAF'S BOOK ON WOMEN'S ANATOMY APPEARED IN MARCH 1672, a hectic, charged time in his own life and in the life of his country. He knew he had produced a masterpiece. Three months later, in June, he married. His wife became pregnant almost at once. Also in June 1672, war broke out, and France invaded Holland. Deeply troubled by news of battles and riots, de Graaf wrote a letter to the Royal Society in July 1672, lamenting "the disaster falling upon the whole of my country." (One month later, a Dutch mob would kill the prime minister and his brother, hang the bodies upside down, and mutilate the corpses.)

In the meantime, de Graaf's onetime friend Swammerdam shoved his way into the picture. First, he wrote an angry letter attacking de Graaf's book and charging him with intellectual theft. Credit for discovering the egg belonged to Swammerdam himself, he insisted, as well as to several others. Swammerdam followed up his letter with a full-length book of his own, *The Miracle of Nature, or the Structure of the Female Uterus.* This work, which appeared only weeks after de

Graaf's, spelled out Swammerdam's views on conception and made his attack on de Graaf public and impossible to miss.

In March 1673, Maria de Graaf gave birth to a son. The parents named the boy Frederick. He died at the age of one month. In August of the same year, de Graaf himself died, perhaps of plague. He was thirty-two.

In the last year or two of his brief life, de Graaf had written a momentous book. Just four months before his death, in the spring of 1673, he had written a momentous letter. De Graaf informed the secretary of the Royal Society that he wanted to introduce a fellow resident of Delft. "I am writing to tell you that a certain most ingenious person here, named Leeuwenhoek, has devised microscopes which far surpass those we have hitherto seen."

De Graaf himself had peeked through one of those microscopes, and he believed he knew what they would reveal.

He did not.

TEN

A WORLD IN
A DROP OF WATER

ANTONY VAN LEEUWENHOEK MADE AN UNLIKELY EXPLORER. Proud and prickly, he was a forty-year-old merchant who sold fabric, buttons, and ribbons to the prosperous citizens of Delft. Leeuwenhoek had no scientific training and only a middling education, but he had infinite patience and far-ranging, unpredictable curiosity. And he had stumbled on a world that no one else had ever seen. More than that, he found himself alone and astonished in a world that no one had ever even *imagined*.

In September 1674, a year after de Graaf's letter of introduction, Leeuwenhoek sent a long letter to the Royal Society. He had written before, reporting on familiar objects like a bee's stinger seen in close-up. (These were the peeks he had offered de Graaf.) This was different. About two hours from his home, Leeuwenhoek explained, was a large, murky lake. For no particular reason, he had collected a vial of greenish, goopy water from near the shore. The next day he examined a drop of water with his microscope. To his astonishment, "very many little animalcules" swam into view.

Leeuwenhoek had seen fleas and mites and other tiny bugs. So had everyone else. But what a dog was to a flea, in size, a flea was to one of Leeuwenhoek's "animalcules." (Leeuwenhoek wrote in Dutch, and "animalcules" represented one early translator's best try at describing his discovery in English. Other translators preferred the term "little animals.") No one had ever suspected that the scale of life continued downward beyond what the eye could see. Why would it, if the world was made for the benefit of humans?

Leeuwenhoek stared in giddy astonishment. "The motion of most of the animalcules in the water was so swift and so various, upwards, downwards, and round about, that 'twas wonderful to see. And I judge that some of these little creatures were above a thousand times smaller than the smallest ones I have ever yet seen upon the rind of cheese in wheat flour, mold, and the like."

Leeuwenhoek was not the first man to peer through a microscope. Half a century before, Galileo had thrilled at the sight of "flies which looked as big as a lamb" and nonchalantly walked upside down. A decade before Leeuwenhoek, Robert Hooke, one of the most talented men at the Royal Society, had examined fleas and slices of cork and dots of ink printed on a page. Hooke marveled at the surprises that his lenses revealed. The perfect, polished tip of a needle turned out to be jagged and irregular, "like an Iron bar [worn] by Rust and Length of Time." The humble flea was outfitted with a perfect "suit of sable Armour, neatly jointed."

But those ventures were only a warm-up act for Leeuwenhoek. Pioneers such as Hooke had taken the known world and, essentially, held a magnifying glass up to familiar scenes. With lenses far more powerful than any that had come before, Leeuwenhoek spied the coastline of a new, undreamed-of continent and then plunged ashore and into the forest, where every leaf on every tree bustled with bizarre forms of life.

Leeuwenhoek confirmed almost at once that the creatures he had found were not unique to the lake where he had gathered them. In a sample of rainwater, he found "little animals more than a thousand times less than the eye of a full-grown louse." These were not mere dots fixed in place but darting, hovering creatures overflowing with energy. Some of the tiny animals "twirled themselves round with a swiftness such as you see in a top a-spinning before your eyes."

In a frenzy—his fascination would endure for fifty years—Leeuwenhoek set out to put the whole world under his lens. Every observation was brand-new, and the Dutch cloth merchant found himself struggling to pin names to countless unfamiliar sights, like Adam in Eden. His approach was utterly unsystematic, zigzagging from project to project as curiosity and happenstance led him.

He did not turn at once to the mysteries of conception, but it was almost certain that he would arrive there before long. Soon after his letter about the bee's stinger, the secretary of the Royal Society had written to him with a request: Would Leeuwenhoek please use his microscope to examine "saliva, chyle, sweat, etc."?

Leeuwenhoek cringed. Saliva and sweat were not problems, nor was chyle, which was food that had been transformed in the digestion process. The problem was the "etc.," which Leeuwenhoek took to be a euphemism for semen. "I felt averse from making further inquiries," he wrote later, "and still more so from writing about them. I did nothing more at that time."

Nothing more to investigate semen, he meant. But nearly everything else in the world came in for Leeuwenhoek's obsessive scrutiny. One of his house servants had the task of collecting fleas for her master to study. He nagged her for blood samples, too, and then he compared her blood with his own and with that from every other creature that darted or crawled into view. He pestered shopkeepers for spoiled bits of food that might harbor tiny pests, and he asked his neighbors to bring him stubble from their beards after they had visited the barber.

Soon after his rainwater experiment, Leeuwenhoek found himself wondering what gives pepper its sharp flavor. Could it be that pepper grains have miniature thorns that stab the tongue? He soaked some pepper in a bowl of water, to soften it. Then, for some reason, he examined the water. He saw *four* sorts of "little animals," and the smallest were "so small that I judged that even if 100 of these very wee animals lay stretched out one against another, they could not reach to the length of a grain of coarse sand; and if this be true, then ten hundred thousand of these living creatures could scarce equal the bulk of a coarse sand grain."

The pepper mystery lay neglected. Leeuwenhoek grappled with a revelation—there was nothing special about a lake or a cup of rainwater. Tiny, hidden life seemed to turn up *everywhere*, not only in one or two rare locales. Leeuwenhoek had not only discovered a handful of new creatures living in an unsuspected world; he had found microworlds all around, all teeming with swimming, tumbling, too-tiny-to-notice forms of life. What were they for? Why had God made them?

STARTLED BY LEEUWENHOEK'S REPORTS BUT UNSURE WHETHER to trust them, the Royal Society set out to look for itself. ("We had such stories written us from Holland," recalled the philosopher John Locke, a Royal Society member, "and laughed at them.") The job fell to Robert Hooke. History has neglected Hooke a bit—he had the misfortune to live in the shadow of Isaac Newton, and no man could have had a more talented or bad-tempered rival—but Hooke was brilliant. His skills spanned so many fields that one biographer dubbed him "England's Leonardo."

An architect and astronomer and engineer, for starters, Hooke could design anything from a cathedral to a mousetrap. More than that, he knew all about lenses. He had written (and illustrated himself) a best-selling book called *Micrographia*, which showed gorgeous drawings of such wonders of nature as a fly's eye, enormously magnified. No man could have been better suited to check Leeuwenhoek's astounding claims.

Lenses themselves were nothing new. Magnifying glasses had been known in the ancient world, and eyeglasses since around 1300. For generations, the inability to see fine detail had been a problem that troubled only a handful of people—seamstresses, goldsmiths, monks

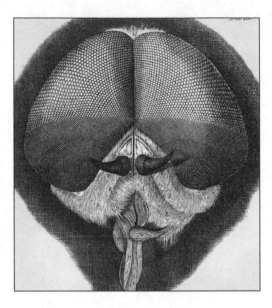

FIGURE 10.1. Hooke marveled at the intricate design of a fly's eye. His drawing showed "near 14,000" eye parts arranged "in very lovely rows."

copying manuscripts or scholars studying them.* But the invention of the printing press, in 1453, brought an explosion in the number of books. For the first time vast numbers of people realized that their vision was inadequate. That led to a flurry of experimentation with different lenses, which led first to eyeglasses and then, as an almost inevitable spinoff, to telescopes and microscopes. In hindsight the path from reading small print to exploring the heavens and the microworld was almost a straight line.

Leeuwenhoek's first encounter with lenses most likely involved examining fabric samples through a magnifying glass. But that ancient invention, and even spectacles, paled next to the seventeenth century's technological marvels. Spectacles and magnifying glasses enhanced a view that you could begin to make out with the unaided eye; the

*The first eyeglasses were reading glasses, for close-up work. Middle-aged users snatched them up, thrilled that their involuntary retirement had suddenly ended. This invention, which may seem small to us, in fact marks a chasm between everyday life as it had always been and as it would become. Until recent times, disabilities and injuries accumulated through the years, never to be mended. Bad eyes, bad ears, throbbing teeth, ripped tendons were all but universal. There was no treatment but fortitude. Eyeglasses hinted at a better future.

telescope and microscope brought into focus sights whose *existence* had never been suspected. The difference was akin to that between a pencil and a magic wand.

On November 1, 1677, Hooke made his first try at seeing Leeuwenhoek's "animalcules." Leeuwenhoek, always secretive about his techniques, had not offered much help. Rather than reveal how he had fashioned his microscopes, he had mailed along testimonials praising his lenses. Eight prominent citizens, including two ministers, explained that Leeuwenhoek had let them peek through his microscopes; they, too, had seen the miracles he described.

That would never do. Hooke set to work, presenting an array of contrivances fashioned from glass and brass. He had taken particular trouble with a set of glass tubes that ranged in size from ten times bigger than a human hair to ten times smaller. Hooke had a theory that a drop of water inside a glass tube might reveal more when examined under a microscope than a drop of water sitting in the open. Hooke labored, the Society members stared, but . . . nothing. "No discovery made."

One week later, on November 7, Hooke tried again. He had found a way to make even narrower tubes, and he had changed the design of the microscope so that it admitted more light. And he had made an extra-strong concoction of pepper-water, by leaving the pepper grains soaking for a full three days. "Notwithstanding the microscope was much better than that shown at the last meeting," the Royal Society's records declared, "yet nothing of Mr. Leeuwenhoek's animals could be seen." Perhaps the excitable Dutchman had somehow gone wrong?

By his third try, on November 15, Hooke had finally triumphed. He had tinkered and fussed throughout the week, and he had let the pepper steep for ten days. In the days leading up to the meeting, he had repeatedly seen the animalcules, and now he showed everyone. There they were! "There could be no fallacy," the Royal Society agreed. "They were near an hundred thousand times less than a mite," and they made "all manner of motions to and fro in the water, and by all who saw them, they were verily believed to be animals."

The records of the Royal Society carefully noted the names of the eyewitnesses who saw this marvelous sight. Christopher Wren, who had designed St. Paul's Cathedral and done pathbreaking work in astronomy as well, headed the list. In modern terms, it would be as if the National Science Foundation had reported finding dinosaurs living on a remote island in the Pacific and had recruited Stephen Hawking and other luminaries to vouch for the discovery.

THE ROYAL SOCIETY NOW BELIEVED LEEUWENHOEK, BUT NO one, not even Hooke, could match his wizardry with a microscope. Leeuwenhoek raced ahead of the pack, with no one else truly even in second place. In long, chatty reports that careened from topic to topic, Leeuwenhoek detailed his journeys through the microcosmos. To a hasty reader, his letters might have looked like the obsessive screeds that turn up on every editor's desk, but it took only a few minutes to see that these notes were thick with revelations.

Leeuwenhoek wrote in Dutch, the only language he knew. This counted as a strike against him, as if a correspondent today sent learned journals handwritten articles rather than electronic documents. (For seventeenth-century readers, Leeuwenhoek's letters had a distinctly down-market tone. "Testicles," for instance, were *zaad-ballen*, "seed-balls.") All proper scholars wrote in Latin. Unable to read their work, Leeuwenhoek found himself confined inside a Dutch-speaking bubble.

The independence may have suited him. Eager for praise but wary of collaborators, he went his own way.* In mingled fascination and

---

*Frustratingly little is known about Leeuwenhoek's personal life. We know, for instance, that both Leeuwenhoek and Johannes Vermeer were born in Delft in October 1632 (their baptisms are recorded on the same page in the records of Delft's *Oude Kerk*), and we know they lived only 150 yards from one another. But small as Delft was, there is no proof that the two geniuses ever met. This would have been a remarkable summit meeting: few people ever had a more passionate interest in lenses and light.

frustration, the Royal Society followed his higgledy-piggledy progress. *Everything* intrigued Leeuwenhoek. He told the Royal Society about a spider with hairs that "stood as thick upon his Carcass as the Bristles on a Hog's back." He nearly blinded himself trying to see what happens when gunpowder explodes. He cajoled whalers newly returned to port into providing him blubber samples. (The flesh "did not smell at all good, because it was rather tainted," but Leeuwenhoek was not a squeamish man.)

He watched attentively as "an hungry Lowse" perched on his hand "to observe her drawing blood." Family and servants pitched in. Perhaps they had little choice. When Leeuwenhoek grew curious about silkworm eggs and how they hatched, he recruited his wife, Cornelia, to help keep the eggs warm. For six weeks, she gamely traipsed about with a small box of silkworm eggs "carried in her Bosom night and day." For another investigation, mother and daughter stood with mouths open wide while Leeuwenhoek poked between their teeth with a toothpick, in search of samples.

He had already found the inside of the mouth to be an especially rich hunting ground. He began by scraping his own teeth, which looked white and clean to the naked eye. Under the microscope, he observed "a little white matter" that appeared "about as thick as batter." Within that plaque Leeuwenhoek saw countless tiny animals "which moved very prettily." Some of the creatures shot along "like a pike through the water." This was a momentous find, one of the first observations of living bacteria.

Leeuwenhoek delighted in showing such wonders to visitors, though some winced while their host beamed. One nervous group rinsed their mouths with vinegar and several rounds of water, to make certain that Leeuwenhoek would not find creatures living on their teeth. To no avail, since vinegar itself abounded with tiny, swimming worms. "Some

---

When Vermeer died, Leeuwenhoek was named executor of his estate. Vermeer had made a middling success of his career but had died bankrupt. To pay off a debt to the baker, Vermeer's widow gave him *Lady Writing a Letter with Her Maid* and *The Guitar Player*. Today their values would be in the tens of millions, or higher. Vermeer owed the baker a bit under $80 in today's money.

of them were so disgusted at what they saw," Leeuwenhoek purred, "that they resolved never to take vinegar again."

Always happy to launch into a frenzy of arithmetic, Leeuwenhoek started calculating. Despite his own clean habits, he noted, "there are not living in our United Netherlands so many people as I carry living animals in my mouth this very day." (His estimates of how many creatures he observed, and their tiny dimensions, were remarkably accurate, although they relied on homemade and improvised measuring sticks. Leeuwenhoek gauged the diameter of a fly's blood vessels, for instance, by comparing them with the number of "divisions [on a brass ruler] that one of the thickest hairs from my beard will cover.")

Many of the challenges he took on sound like the labors of Hercules, if Hercules had been the size of Tom Thumb. "I also with much trouble took out the testicles of a flea, and placed them before my microscope," Leeuwenhoek told the Royal Society. A few sentences later, he noted that "I took also some flesh from the feet of a gnat." He thrilled at the sight of blood in the capillaries—Leeuwenhoek was the first to see capillaries and the first to see blood cells—and he stared entranced at a spectacle that Harvey had only imagined. In the tiny blood vessels in a tadpole's tail, Leeuwenhoek saw the blood traveling as clearly "as when, with the naked eye, we see the water leaping high out of a fountain and then falling down again."

He wrote with unquenchable excitement, a tour guide who grew ever more enthusiastic as the samples he collected grew less and less enticing. Blackheads intrigued him, and so did earwax and pus and his own diarrhea (where he found "animalcules a-moving very prettily," with long bodies and flat bellies, and "furnished with sundry little paws").

Curious to see what a callus looked like in close-up, Leeuwenhoek stripped off his socks and shoes. Then he placed his foot upon a piece of blue paper and recruited his servant. The hapless man, "partly with his Nails and partly with a Penknife," managed to pry off a bit of skin. Intrigued with his first glimpse of this "hard Skin," Leeuwenhoek recruited a mason, then a carpenter, and then a plowman, and sliced skin samples from their work-hardened hands.

In these early days of science, devout and brilliant thinkers like Kepler, Galileo, and Newton found inspiration in contemplating the starry tapestry that God had flung across the heavens. Antony van Leeuwenhoek, a titan of another stripe, followed a different path. To grasp the magnificence of God's creation, he fervently believed, one could study not the fiery sun and the stately procession of the planets but pond scum, and spit, and blood, and fleas, and worms.

ELEVEN

"ANIMALS OF THE SEMEN"

O N AN AUTUMN NIGHT IN 1677, LEEUWENHOEK AND HIS WIFE
made love. He leapt up "immediately after ejaculation before
six beats of the pulse had intervened," and ran to his microscope
with a sample of semen. There Leeuwenhoek saw "so great a number of living animalcules that sometimes more than a thousand were
moving about in an amount of material the size of a grain of sand."
He did not tell the Royal Society if Mrs. Leeuwenhoek found this a
thrilling discovery.

This observation would eventually take its place as a towering milestone in the history of science. But Leeuwenhoek, who ordinarily was
happy to trumpet his finds, played this one down. He reminded the
Royal Society that this had been *their* idea, not his. With uncharacter-
istic self-consciousness, he noted that he had obtained his test sample
by "conjugal coitus" rather than by "sinfully defiling myself." He even
took the trouble of having his letter translated into Latin, perhaps to
shield tender readers. Despite his precautions, Leeuwenhoek wrote,
he recognized that his observations might "disgust or scandalize the

learned." He gave the Royal Society permission to publish his letter or burn it, as they saw fit.

There is no mistaking what Leeuwenhoek saw on that long-ago evening when he dashed out of bed. The tiny creatures under his microscope "were furnished with a thin tail, about five or six times as long as the body." They corkscrewed along "owing to the motion of their tails like that of a snake or an eel swimming in water," as if they were racing toward some important destination.

There is no mistaking what he saw, but Leeuwenhoek did mistake it. Historians hail Leeuwenhoek as the first man to see sperm cells, but that is only partly right. He did see tiny cells swimming in his semen, but he did not suspect that they had anything to do with procreation. Instead, he thought he had found micro-animals that happened to live in semen. After all, hordes of microscopic creatures seemed to cavort everywhere—in drops of water, in tree sap, on your teeth, between your toes. Why shouldn't semen have creatures of its own?

For half a dozen years, Leeuwenhoek held to this view. Like Columbus, who "discovered" the new world while insisting that he had found the Indies, Leeuwenhoek vehemently misunderstood his own discovery. And just as Columbus immortalized his error in the name "Indians," so Leeuwenhoek called sperm cells "little animals."

Most scientists agreed with him. Odder still, they *continued* to believe that sperm cells were tiny animals that had nothing to do with sex or reproduction long after Leeuwenhoek himself had changed his mind. A century and a half after Leeuwenhoek first saw sperm cells, well into the 1800s, the animal theory was still conventional wisdom. Meticulous illustrations in biology journals showed carefully drawn sperm cells, with neatly labeled mouths and bladders and other organs, or showed sperm cells and tapeworms side by side, by way of comparing different microscopic animals.

In 1830, an article in the *Lancet*, the leading medical journal of the day, included spermatozoa in a discussion of intestinal worms. "Animalcules are always found in the semen, which is, no doubt,

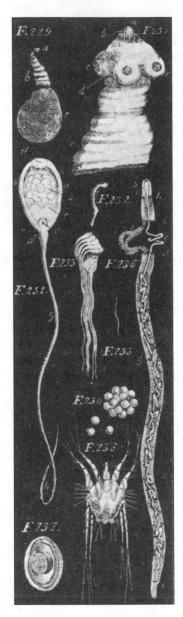

FIGURE 11.1. An illustration depicting a variety of parasites, from a medical text written in 1840, shows a sperm cell (at left) along with tapeworms and other unappealing creatures.

their natural habitation," the *Lancet* noted. "They cause no inconvenience and, no doubt, serve some important purpose." Even the word "spermatozoa" reflects this long-lived error. The word was coined in the year 1827 by a scientist who believed that sperm cells were worm-like creatures from the genus *Cercariae*. "Spermatozoa" means "animals of the semen."

Some scientists found a different way to miss the point of Leeuwenhoek's discovery. Sperm cells were not little animals; they were stirring rods! This view, too, lasted until the 1800s. The idea was that semen was important but sperm cells were not, except as a means of keeping the semen from stagnating. Biologists looked at semen, just sitting there waiting for the big day, and contrasted it with blood, which was in constant movement. They saw the difference between a river's vitality and a pond's torpor. What could be plainer than that Leeuwenhoek's wriggling, twisting, perpetually swimming animalcules were animated swizzle sticks?

These were colossal errors. With an essential clue to the babies mystery in their hands, scientists had fumbled it away. It was not only that they had picked up a smoking gun and failed to see what it was. Worse, they had watched the wisps of smoke rise into the air and said, "What a curious tea-kettle!" and then set the gun aside.

HOW DID EVERYONE GET IT SO WRONG? PART OF THE PROBLEM was that Leeuwenhoek and his peers had no vocabulary that fit his discovery. Today we talk about cells as the building blocks of life. Cells are to biology what atoms are to chemistry, and the first page of every biology textbook explains that all living organisms are composed of cells. We talk of sperm cells and egg cells as if they are natural features of the living world that no one could miss, like trees and birds. But the cell theory dates from the early 1800s, well over a century after Leeuwenhoek. (I have sometimes referred to "sperm cells" because the term is almost unavoidable, but it is an anachronism.)

In the 1600s and 1700s, when scientists saw anything little that looked alive, they took for granted that it was some sort of bug or beast or worm. So when Leeuwenhoek looked at semen through his microscope and saw "animalcules" swimming purposefully along—not drifting along like sticks in a stream or floating haphazardly like seaweed but thrashing their tails and racing—he immediately classified them as animals. What else could they be?

(Eggs, if they had been found first, might not have led to the same wrong turn. Early scientists thought of eggs as immobile and associated them with calm and repose. Sperm cells, with their thrashing tails and vast numbers, called to mind tumult and confusion and, above all, activity. In short, life.)

Even so, there were problems with the notion of spermatozoa as microtadpoles that happened to have found a pond in the testicles. The biggest riddle was simply put: If these were animals, where had they come from? They had landed in a most unlikely home (*"Out of the way location, no central heating, poor views"*), and they could hardly have drifted in on the breeze or hitched a ride on a bite of food. Who were these interlopers?

Eventually this question would take on tremendous importance. In the meantime, it was put aside in favor of more immediate chores, like sorting out whether animalcules turned up in all sorts of male animals and at all stages of life.

L EEUWENHOEK HIMSELF BELIEVED HE HAD SPOTTED SOMETHING in semen far more significant than tiny, swimming animals. *This* observation, though in fact it was utterly mistaken, filled him with "wonder" about the mystery of procreation; the sight of his tiny animals, which genuinely was a breakthrough for the ages, had stirred no such thoughts. Within the liquid part of the semen—not inside the swimming animalcules—Leeuwenhoek reported seeing "all manner of great and small vessels." He had "not the least doubt," he told the Royal Society in November 1677, "that they are nerves, arteries, and veins."

Convinced that he had made a giant find, Leeuwenhoek returned to the same theme in another letter to the Royal Society a few months later. The nerves and blood vessels he had seen, he proposed, somehow gave rise to all the parts in a developing embryo. "It is exclusively the male semen that forms the fetus and all that the woman may contribute only serves to receive the semen and feed it."

This was bizarre for several reasons. For one thing, Leeuwenhoek had made a momentous discovery and then ignored it. For another, the vessels that filled him with wonder do not exist, and no one has ever figured out just what it was that this brilliant observer actually saw. Finally, Leeuwenhoek's assertion that semen was vitally important to conception and that the egg played no role whatever came out of the blue. He dismissed the egg contemptuously but did not bother to provide any arguments to back his claim.

The secretary of the Royal Society, a physician and botanist named Nehemiah Grew, didn't buy it. He sent Leeuwenhoek a challenging letter. "Our Harvey and your de Graaf"—the eminent Englishman and the accomplished Dutchman—had proposed a strikingly different picture of conception. Those renowned scientists had focused almost entirely on the egg, Grew reminded Leeuwenhoek. In their view, semen played a distinctly secondary role; it merely started the egg on its path toward development, and it did so in the most airy, ethereal fashion. Semen did not physically contact the egg, wrote Grew, but merely awoke it with "a certain breath" from a discreet distance. How did that picture of an ever-so-gentle air kiss fit with Leeuwenhoek's confused account of tangled nerves and arteries and swimming eels racing one another?

Always cantankerous when challenged, Leeuwenhoek lashed out. He informed the Royal Society that he had heard of one author who had cited *seventy* scientists who echoed Harvey and de Graaf. *What of it?* If there had been "seventy times seventy more" all chanting in unison, still he would "maintain that every one of them has erred."

But Leeuwenhoek did more than sputter in irritation. Spurred by Grew's challenge, he looked closely at just what de Graaf and Harvey had written about conception. By this roundabout path, finally, he came to recognize the importance of the tiny animals he had seen swimming in his semen seven years before.

The crucial experiment came on December 31, 1684. At eight o'clock that morning, and then again at two in the afternoon, Leeuwenhoek watched as a male dog mated with a female in heat. Then he had the female killed (with an awl thrust into her spine) and cut open. With his naked eye, he saw no sign of semen. So far this was exactly what Harvey and de Graaf had reported. (From our vantage point, the failure to find semen makes sense. Semen would have been nearly impossible to see because it would have dispersed in the wet mess of a freshly opened body.)

Now Leeuwenhoek looked through his microscope. Neither Harvey nor de Graaf had done such a thing. "I discovered to my great satisfaction an abundance of living animalcules, these being the male seed of the dog," he wrote triumphantly. The number of tiny animals was so huge that "in my estimation an odd hundred million or so would make little difference to the reckoning."

Leeuwenhoek had earlier proclaimed what he believed was the true story of conception. At the time he'd had no proof. Now, he felt, he had proved his case so decisively that only the "obstinately opinionated" could deny it. The dog he had cut open was not the point; the dog was the arrow that showed the way to a general truth. "A human being originates not from an egg but from an animalcule that is found in male sperm."

From the start, Leeuwenhoek had hailed semen as all-important. That did not change. What was new here was that Leeuwenhoek had moved the spotlight. Now he made no mention of the mysterious "vessels" and nerves that had once struck him as the key to the procreation riddle. Instead, he focused on the millions upon millions of tiny, swimming animals within the semen. *Do you see? Look there!* The secret of life lay hidden within those microscopic, wriggling bodies.

With the animalcules now in a starring role, Leeuwenhoek banished the other player in the drama altogether. Despite what Harvey and de Graaf had insisted, the egg had no role in conception. In Leeuwenhoek's account, the male delivered the animalcules to the female, where they burrowed into the uterus, which nourished them.

Leeuwenhoek's picture drew heavily on the old analogy between a man's semen and a tree's seeds. Animalcules grew into animals, he explained, just as apple seeds grew into apple trees. The analogy was based on a deep misunderstanding. The seed of a plant is an embryo, a product of sexual fertilization, rather than just a male sex cell. But scientists did not start to sort out the riddles of plant sex until around 1700. (Everyone knew that plants grow from seeds, but nobody understood where those seeds came from.) Leeuwenhoek had scurried by other trouble spots, too. He simply declared that the animalcule gave rise to the embryo, for instance, without explaining how in the world that could happen. Nor did he explain why a man produced millions upon millions of sperm cells, if a single one would have sufficed.

Instead, he devoted his energy to shooting down his rivals. They claimed that the egg made its way from the ovaries to the womb. Leeuwenhoek demanded to know how that happened. Were we to believe that somehow an "egg was sucked from the egg-nest" by the floppy Fallopian tubes, like a sailor snatched from the deck of Odysseus's ship by some long-tentacled sea monster?

More ludicrous still, de Graaf's supposed eggs were large (as we have seen, de Graaf had confused egg follicles with the much smaller eggs within them), while the tubes they purportedly passed through were small. How did *that* work? And if eggs played such a key role in conception, why was it that Leeuwenhoek saw no sign of them when he examined female dogs with his microscope? After all, he had discovered sperm cells, which were far tinier. In December 1684 and again in January, Leeuwenhoek examined the Fallopian tubes of dogs who had mated moments before. He saw nothing noteworthy except a few "globules" that were certainly not eggs. "Had there been one particle in them no bigger in size than the hundredth of a grain of sand, I doubt not but I should have found it."

Why he did not is a genuine mystery. Leeuwenhoek's honesty is beyond question. By fluke or bad fortune, the eggs he might have seen somehow escaped him. Perhaps he had inadvertently dislodged them in the course of his probing.

Never a man for diplomatic phrase making, Leeuwenhoek denounced the egg theory as "addle-pated," "fantastic," and "entirely erroneous." He had looked in "so-called ovaries" and he could report that the "so-called eggs" they supposedly contained did not exist. So Leeuwenhoek proclaimed, and so he would argue for the rest of his long life.

Better still, he told the Royal Society in 1685, he might have seen something that cracked the conception mystery wide open. "I have sometimes imagined, as I examined the animalcules in the male seed of an animal, that I might be able to say, there lies the head, and there, again, lie the shoulders, and there the hips." Straining to make out tiny details at the limits of human perception, Leeuwenhoek was sincere but mistaken.

He immediately added that he was far from sure that he had truly seen this marvelous sight. "Not having been able to judge of this with the slightest degree of certainty, I shall not, therefore, affirm this as definite, but rather hope that we may, at some time, have the good fortune to come across an animal whose male seeds will be so large that we can recognize in it the figure of the creature from which it has come."

He continued searching. Fifteen years later, on Christmas day in 1700, he wrote to the Royal Society about the animalcules he had seen in a ram's semen. "The parts lying in such an animalcule do not resemble a lamb, he conceded, "yet the parts lying therein may in a short time assume the shape of a lamb when they have received nourishment in the womb."

Leeuwenhoek pressed on, certain that the future animal had somehow to be concealed inside the sperm cell. It was a question of logic. How could a tree sprout branches if the branches were not somehow in the seed already? *Something* could not materialize from nothing.

He vowed to look harder.

⸎ PART THREE ⸎

RUSSIAN DOLLS

"Very dangerous things, theories."

—DOROTHY SAYERS,
The Unpleasantness at the Bellona Club

TWELVE

DOLLS WITHIN DOLLS

T HE STAGE SEEMED SET FOR A GIANT ADVANCE. THE ANATOMISTS, with de Graaf in the lead, had made their breakthrough—women had eggs; *that* was their role in conception. The microscopists, following Leeuwenhoek, had their own triumph to celebrate—the "tiny animals" that swam in semen, and not the semen itself, were crucial. Thrilled at the discovery of those animalcules, Leeuwenhoek's camp scarcely had time for anyone's talk of eggs.

The next task, one might have guessed, would be for the two sides to make a truce and come to some agreement on just what roles egg and sperm played. It didn't work out that way. Instead, scientists split into two warring camps, ovists and spermists, each dedicated to the proposition that *their* side was truly vital, while the other was necessary, perhaps, but distinctly secondary.

Worse still, the two sides did manage to come together on one key issue, but that one instance of agreement led not to progress but to a crucial wrong turn. That mistake was an astounding theory called preexistence, which would entrap scientists for generations.

The problem began with the battles between ovists and spermists. We have seen already how Leeuwenhoek sneered at his "addle-pated" rivals for their devotion to the egg. Ovists lobbed back insults of their own. Such name-calling was far more than the customary squabbling of competitors scrambling for recognition. The two sides had made the crucial mistake of thinking that they could get along without one another.

Ovists insisted that the embryo lay concealed within a woman's egg. It had been there all along, before any man had ever come into the picture. To hear the ovists tell it, semen was simply the key that set a living clockwork in motion. Spermists maintained just as fervently that the embryo lay concealed within the sperm. It had always been there, waiting. When a couple have sex, the man's sperm carries the embryo to his partner, where it grows. Either there is no egg at all (this was Leeuwenhoek's view), or egg and womb are merely food and incubator for the new arrival.

The striking feature, to modern eyes, is that neither side granted the other any role in forming the embryo in the first place. Instead, they proposed a remarkable alternative: parents do *not* form their children. On the contrary, those children were preformed before their parents ever laid eyes on one another. On this point, and only on this point, both sides agreed. It was God who had done the preforming, and according to the most popular version of this theory, he had done so at the dawn of time.

This was a misguided, not to say bizarre, theory. But it is crucial to note that this was mainstream science, not a fringe doctrine, and Europe's most eminent thinkers all signed onto it. The preexistence theory, as it was called, would reign for more than a hundred years, from late in the 1600s to the very end of the 1700s. In hindsight, a hundred-year reign of error is a colossal blunder, but in all those years most scientists had no inkling that they had veered off course.

At the time, the new theory seemed all but irresistible. Scientists embraced it so ardently partly because they had come to scorn a belief that had satisfied all their predecessors since the Greeks. From Aristotle to Harvey, scientists had taken for granted that some intangible force directed the embryo's development; this vital spirit, though nearly impossible to describe, guided living plants and animals in a myriad of ways. This life force directed plants to send out shoots and roots, broken bones to mend, and wounds to heal. Above all, this animating spirit directed embryos to take on their destined shape.

In the new, mechanically minded age, that notion had come to seem alarmingly vague, perhaps even empty. To say that the embryo somehow "knows" how to grow from a featureless blob into a full-fledged organism with wings and a beak or with fingers and toes was to invoke magical or occult forces. That was less an explanation than a shout of "Abracadabra!"

Who would cling to such an outmoded creed when they could opt for a scientific, hard-headed alternative?

THE PATH TO THE NEW THEORY WAS A CURIOUS ONE. IT RELIED partly on experiment and observation, as you might guess, but far more heavily on faith. That faith took various forms: faith in the power of reason, faith in the orderliness of nature, and, above all, faith in God. Armed with unshakable religious faith, physicists in the Age of Science had explained the workings of the heavens. Armed with unshakable religious faith, biologists in the Age of Science raced into a swamp where they wandered, lost and frustrated, for a century.

Religion guided physicists to discovery after discovery because it seemed that God truly is a mathematician. No one has ever explained why, but for proof you had only to look up. Take Halley's comet, for instance. For countless millennia, on a once-every-seventy-five-years timetable, it had traced an enormous ellipse across the heavens. Throughout nearly all that expanse of time, no human being had ever imagined such a thing as an ellipse. Finally, in 1705, Edmond Halley proved that the comet now named after him had been tracing perfect ovals all through the numberless generations when human beings still cowered in caves and struggled to count on their fingers.

It's easy to imagine a universe where things happen any which way. The sun and stars would flicker and blink randomly, and comets and planets would travel not in smooth, perfect curves but in a drunkard's zigzagging lurches. But that is not our world. Falling rocks obey mathematical law. So do crumpled pieces of paper tossed toward a wastebasket. Rainbows form precise arcs. The ripples that spread across a pond when a duck lands on the water move outward in perfect circles, at rates governed by specific laws, and the sounds of the duck's quacking obey yet another set of concise mathematical laws.

The physicist Eugene Wigner, a Nobel laureate, wrote what is still the best attempt to explain why the universe follows such neat rules. But even this landmark essay, from 1960, amounts to an eloquent confession of ignorance. Wigner acknowledged his mystification in his title, "The Unreasonable Effectiveness of Mathematics in the Physical Sciences." He concluded on a humble note. "The miracle of the appropriateness of the language of mathematics for the formulation of the laws of physics is a wonderful gift which we neither understand nor deserve." All physicists share Wigner's feeling of baffled gratitude.

But for the biologists of the late 1600s and 1700s, religion led not to enlightenment but to muddle. For these early scientists, the great, unfathomable mystery was explaining how anything as complicated as a living creature could ever have been formed. They could imagine only two possibilities.

Think of any bit of matter whatsoever, from a boulder to a leopard. Either it had been built according to a plan, or it had arisen by chance. In the case of a living organism, there was no room for doubt. The closer one looked at a living creature, with its countless parts, each exquisite in itself and all of them working together in perfect harmony, the more the designer's hand proclaimed itself. And that infinitely skilled and infinitely patient designer, every scientist recognized, could be none other than the Creator himself.

Not even Isaac Newton, perhaps the most brilliant scientist who ever lived, could imagine that there could be design without a designer.

So mad a notion practically rebutted itself. Suppose for a minute, for the sake of argument, that the universe was composed entirely of tiny billiard balls and that those balls collided according to strict mechanical laws. It was not just ludicrous but blasphemous to think that those careening atoms might spontaneously form a table or a house, let alone a mouse or a mastodon. How could chaos give rise to order?

If the atheists were right, the world would be awash in monsters, misshapen beasts with legs sprouting from their backs or blinking, useless eyes. "Were men and beast made by fortuitous jumblings of atoms," Newton scoffed, "there would be many parts useless in them—here a lump of flesh, there a member too much."

Later thinkers took up the argument. The more details they filled in, the more iron-clad it seemed. Suppose you happened to stub your toe on a stone, wrote the clergyman and naturalist William Paley. You might say, "What of it? Perhaps it had been lying in the grass forever." But suppose you stubbed your toe on a watch. *Notice how the wheels and gears mesh perfectly. Observe how they would not work at all if any of them was even the tiniest bit bigger or smaller. Look at the steel springs and the glass face, each material perfectly suited to its purpose.* In such an event, Paley demanded, who could fail to see that the watch had been designed? And who would deny that a living creature was vastly more complex than the most sophisticated watch?

Nearly two centuries would pass before a shy, sickly, mathematically illiterate Englishman named Charles Darwin explained where Newton and Paley had gone wrong. In the meantime, Leeuwenhoek and all his contemporaries took for granted that God had designed every feature of heaven and earth, down to the last detail. This was intelligent design with a vengeance.

For physicists, the notion of a designer at the controls posed no problem. They had only one creation to explain, and they had the story on the best possible authority. God had created the sun and the starry heavens once and for all. Then he'd set the clockwork running. But for biologists, the problem was enormous. Since every living creature had been designed by an all-knowing, all-powerful Creator, every feature of that design had to be impeccable. That was trouble. How to account, for instance, for babies born blind or missing an arm or for twins joined at the head?

For Darwin, such tragedies had a straightforward explanation: in the tricky business of development, a lot can go wrong. Suppose, though, you took as an unchallengeable premise that God's design was perfect in every feature. Then what?

And not only was the living world marred by mishap and tragedy. Even when things ticked along according to plan, life on earth was not only disorderly—in utter contrast with the harmony that marked the cosmos—but teeming with ugly and badly behaved creatures. Why had the Creator fashioned tapeworms and rats and fleas? What of the poor lamb, torn open by wolves?* The best of all possible designers, though indisputably a master of technique, had a hard-to-deny quirky streak. Couldn't the camel have been a touch handsomer? What good were houseflies? Was there really an urgent need for hundreds of thousands of species of beetle?†

Before Darwin, these were baffling questions. Afterward, in the dog-eat-dog world that evolution would one day describe, it was no surprise that some dogs thrived and many faltered (and that cats had to watch out). But none of that made sense in the 1600s and 1700s, when the natural world was supposedly not a scrum but a waltz.

Far more important and more perplexing, though, was *this* riddle: What was God's role, today, in the creation of life? The Bible spelled out the details of the original creation—God had created plants and trees on the third day, birds and fish on the fifth, and humans and land animals on the sixth—but new life popped into existence every minute of every day, *right now*. Did God watch over the first appearance of every new plant, animal, and human being?

---

*The hazards of the inanimate world, like earthquakes and volcanoes, did not pose any theological riddles. God had flooded his creation, after all, and everyone understood that at any moment he might punish sinning humankind once more.

†When a theologian asked J. B. S. Haldane, one of the eminent figures in twentieth-century biology, what we can learn about the mind of God by studying his creation, Haldane scarcely paused to ponder. "He must have been inordinately fond of beetles."

Think just of human babies. It hardly fit with the seventeenth century's notion of divine dignity to picture God, like a voyeur, looking in on every conception in the world.

But what was the alternative?

THE ALTERNATIVE WAS ASTONISHING. IT WAS, HOWEVER, A MATter of inexorable logic. God, a designer whose way of working was necessarily perfect, did not act in one way here and in a different way there, or in one way today and in a different way yesterday. His divine guidelines spanned the universe. He had created the inanimate world in one go. That was certain. It followed, inevitably, that he had created the *living* world in one go.

It didn't look that way, but that was because people had not looked hard enough. Science taught a different lesson: Every person who would ever live, down to the end of time, had been created at the same time. (And not just every person, though human beings are our focus here, but every living plant and animal of whatever sort.) God had created them all in the beginning, and since that first flurry of activity, there had never again been any *new* creation of life.

That was the only possibility that was in keeping with God's nature. But where had the countless creatures who were created so long ago kept themselves all this while? In particular, what had God done, in the world's first week, with all the humans destined to be born in the 1200s or 1500s or in some century yet to come? He had stashed them away. They waited, like a series of ever-smaller Russian dolls one inside the other, in Adam's testicles or in Eve's ovaries. When the time came, each one would have its turn on stage.*

The picture of worlds within worlds, forever, sounds like a druggy hallucination. But scientists in the seventeenth century managed, more or less, to accept the big picture. They concentrated their energy,

---

*Jan Swammerdam cited yet another argument in support of the theory that all life had begun at once: it explained the essential Christian doctrine of original sin, which holds that every human being is *born* bad, tainted by a sin passed on from Adam.

instead, on fighting over whether it was Adam or Eve who carried that infinite doll collection.

Scientists spoke casually of "encasement" and "preformation" and "preexistence." Today, in much the same way, we attach names like "Big Bang" or "black hole" to impossibly arcane notions, as if the act of labeling provides insight. A French scientist, writing in 1700, tried to color in some details. "In one single spermatic worm, there is an infinity of organized bodies capable of producing fetuses and children, for infinite centuries, always smaller and smaller in relation to one another." Perhaps sensing his reader's perplexity, he added a censorious remark. This picture would "only appear bizarre to those who measure the wonders of the infinite powers of God according to the idea of their senses and of their imagination."

In truth, this was a theory to set nearly anyone's head spinning. In Adam's body were testicles; in those testicles were sperm cells; in those sperm cells were miniature proto-humans; in *their* testicles were micro-miniature proto-humans, who had testicles of their own, within which . . . and so on, forever. (If one of those infinitesimally small proto-humans happened to be female, she might grow up to have children with some man; his sperm cells would contain miniature proto-humans whose sperm cells contained even tinier proto-humans, and the whole process would continue along that new branch of the family tree.)

To modern minds, the theory seems impossibly far-fetched, and biology textbooks for a century have delighted in mocking it. But the mockery is misguided. This silly-sounding theory was espoused by men who were not in the least bit silly. Many of them were brilliant, and all of them were serious. (It is worth noting, too, that nearly everyone in the 1600s believed that the Earth was a mere six thousand years old. That meant that Adam and Eve lived only a few hundred generations ago, which made the sequence of Russian dolls a bit less daunting to contemplate.)*

---

*The Russian doll analogy, which features in every history of biology, captures the strange notion of encasement, which is indeed the crucial feature of the preexistence theory. But the analogy is misleading in one important regard. The model of a doll inside a doll inside a doll would mean that each generation in a given family had only a single member. Instead, each parent doll should have within it one doll for each child.

What would strike a later age as impossibly far-fetched impressed the seventeenth century as grand and inspiring. In a sense, the doctrine was a counterpart to today's string theory, the "theory of everything" that many of the most esteemed figures in physics have been laboring on for decades. Both theories purported to explain nothing less than the nature of reality; both gleamed with intellectual rigor; both took abstraction so far that theorists swooned and experimentalists shuddered.

When physicists today discuss their work, they take for granted that their theories are impossibly arcane. That is not a sign of trouble, as they see it, but merely a fact about the world. Edward Witten is sometimes hailed as the greatest living physicist. In a recent interview, he waved aside objections that string theory is so far from what humans can picture or experiments can test—it bristles with such mysteries as extra dimensions that hide, curled up, so that we cannot see them—that it is more poetry than science. Those complaints were beside the point. "The majesty of string theory" was the point. "The universe wasn't made for our convenience," Witten snapped. In the 1600s, thinkers might have grown dizzy at the notion of an infinite succession of ever-tinier living beings, one within the other, and of those lives unreeling across endless centuries. What of it? Who said that finite man can grasp the ways of his all-wise, all-seeing Creator? The proper response to the theory, one early scientist observed, was "holy awe."

Nor did it count as a mark against the theory of preformation that, even armed with their newfangled microscopes, scientists could not *see* these infinitesimally small dolls. They carried on unfazed, like Einstein and his fellow scientists in the twentieth century who deduced the existence of atoms they could not see. Preformation's great virtue was intellectual, not visual. It did not reveal the world, but it made sense of it. As one naturalist proclaimed, in rhyme:

> We must believe
> What Reason tells: for Reason's piercing Eye
> Discerns those Truths our Senses can't descry.

You cannot explore what you cannot imagine. In the 1600s and 1700s, scientists could not imagine a world without God. Tied to a particular notion of the divine plan, they fashioned an argument whose every link they deemed indisputable. They ran into trouble not because they indulged in wild flights of imagination but because they had such faith in logic that they followed their argument no matter where it led them. Mesmerized by the power of reason, earnest scientists ran alongside Alice and through the looking glass.

IT WAS FAITH IN A PERFECT, ALL-POWERFUL GOD THAT PROVIDED the sturdiest pillar for this new doctrine. But other strong arguments bolstered the same case. Look at a tulip bulb, observed the French philosopher and priest Nicolas Malebranche, in 1672, and you can see the tulip's structures already present in miniature. This is still a classroom exercise today.

A renowned Italian biologist named Marcello Malpighi reported an analogous finding, in 1674, this time in chickens. Like Aristotle and Harvey before him, Malpighi had examined chicken eggs as they developed, day by day. Unlike his predecessors, he had used a microscope rather than his naked eye. Within a new-laid egg, Malpighi informed the Royal Society, he could already discern the embryo's structure. This was exactly in keeping with the doctrine of preformation. Malpighi was one of the great authorities on the microscope, and his judgment carried weight. (He was mistaken, though, because he did not realize that the embryo starts growing *before* the egg is laid. What he took to be the embryo's earliest moment was actually a scene from later on in the tale.)

These reports of tulips and chickens were scattered examples, but scientists' hope and expectation was that they foretold the general story. Like the first crocuses of spring, they carried a message not of flukiness but of bounty soon to come. As microscopes improved, countless new organisms would no doubt unveil their own preformed, tidily bundled structures.

But in the meantime, even before microscopes had gained the power to show those preformed structures, they had revealed sights that made that outcome seem inevitable. Every facet of every creature ever examined, no matter how tiny and elusive, turned out to be exquisitely made. That didn't prove that every blurry, out-of-focus image concealed a sharp image within, but it certainly implied such a conclusion. Jonathan Swift, who was fascinated with both telescopes and microscopes, wrote his famous lines on the flea in 1733:

> So, Nat'ralists observe, a Flea
> Hath smaller Fleas that on him prey,
> And these have smaller Fleas to bite 'em;
> And so proceed ad infinitum.

The words "ad infinitum" were more than a clever rhyme. Swift and his contemporaries took literally the notion that living creatures could grow smaller *forever*. A century later, with the advent of cell theory, biologists would learn that nature comes with size limits. Living organisms are made of cells, it would turn out, and anything alive must be at least one cell big. (We can imagine an infinitely tiny cell, but in practice there can be no such thing because even one-celled organisms must contain certain fixed-size parts, like water molecules.) But in the 1700s, microscopists' astonishing reports seemed to tell a different story.

So did the brilliant success of the newest and most powerful weapon in the physicists' armory. This was calculus, a kind of conceptual microscope that Isaac Newton and his followers used to dissect the workings of the heavenly clockwork. Newton's invention (to his fury, his rival Gottfried Leibniz had made the same breakthrough) was built explicitly on the idea of "infinitesimals," which were infinitely short distances and infinitely brief stretches of time.

Since the days of Zeno and his infuriating paradoxes, about four hundred years BCE, mathematicians and philosophers had been terrified by the notion of infinity. To contemplate infinity was to flirt with madness. The great intellectual breakthrough that Newton and Leibniz achieved late in the 1600s—the breakthrough that had eluded Archimedes and all his successors for nearly two millennia—was finding a way to freeze-frame the moving, changing world by slicing it up into infinitely many still images. Then, with reality pinned down like a slide under a microscope, they could zoom in and examine each scene in close-up.[*]

Archimedes and any of the others might have done it. Mathematics requires no equipment beyond pen and paper, and the Greeks had mastered all the necessary intellectual tools. What they lacked was nerve. Newton and Leibniz had nerve in superabundance. They harnessed infinity and revolutionized the world. For scientists in the 1600s and 1700s, the moral seemed plain: If the key to understanding the physical world was to look in ever greater close-up, why wouldn't exactly the same key unlock the secrets of the living world?

Antony van Leeuwenhoek, far too impatient a character to wander for long in the thickets of philosophy, rarely spoke out on the Russian doll theory. He preferred to focus on making the case for the sperm, and not the egg, as the key factor in conception. Occasionally, though, in brief passages in his letters to the Royal Society, he ventured into the open. In a letter in August 1688, he boiled his argument down to a single axiom, like a line in Euclid: "God, Lord and Omniscient Maker of the Universe, makes no new creatures." That single premise contained implicit within it the whole theory, much as Adam's loins contained the whole history of humankind.

Leeuwenhoek took pains to head off likely criticisms of his spermist views. He warned his fellow scientists that they might find it almost unimaginable that "in an Animalcule from the Male sperm, which is so incredibly small, so great a mystery as a Human body can be enveloped." But there were no limits to the miniature marvels the microscope had

---

[*]Both Newton and Leibniz believed in preformation. Newton was an ovist, Leibniz a spermist.

revealed. Why shouldn't the animalcule theory be true, when Leeu-wenhoek had seen millions upon millions of tiny, complicated crea-tures occupying a space no bigger than a grain of sand?

Yes, it was astonishing, he scolded. *Of course it was!* But what was the point in wailing, "I can't believe it!" Like it or not, it seemed likely that "an animalcule from the male seed of whatever members of the animal kingdom, contains within itself all the limbs and organs which an animal has when it is born." (He never explained just how those embryos would grow, but the idea seems to have been that they would unfold along built-in lines, a bit like those toys for toddlers where a tiny, nondescript bit of sponge dropped into water opens up into a duck or a dinosaur.)

Leeuwenhoek went out of his way to emphasize that he did *not* believe that the proto-human within the sperm cell was a minuscule but perfectly proportioned human being. His claim was only that the proto-human contained all the parts that would one day grow to make a proper human. During the course of development those proportions might change drastically; nearly any sort of stretching and shifting was possible. The point was merely that no *new* parts could appear out of the blue.

In the meantime, new evidence had come along to bolster Leeu wenhoek's case. Just as he had proclaimed, living creatures had been preformed all along.

The evidence came from a most unlikely direction.

THIRTEEN

THE MESSAGE IN GOD'S FINE PRINT

Since earliest times, humans had done their best to ignore the whole buzzing, biting, teeming insect world. Butterflies and bees were fine, but who didn't despise flies and fleas? And then, for scientists at least, the world shifted. Starting late in the 1600s, they focused intently on insects, with fascination and deep respect. These tiny, intricate creatures were important in their own right, scientists explained, but that was the least of it. More important, insects pointed to truths that applied across the living world, all the way to humans.

One reason for the focus on insects was practical: they were easy to find, easy to study, and short-lived. Like fruit flies in modern-day genetics labs, insects made ideal research tools for scientists eager to probe the secrets of sex and growth. But, as so often was the case in the 1600s and 1700s, the deeper reason had to do with religion.

The more that scientists scrutinized insects, the more they found to marvel at, from the brilliant sheen of a beetle's shell to the intricate structure of a fly's compound eyes. They drew what they took to be the inescapable conclusion: here was new insight into God's cosmic design. By the time they were done, they felt confident that they had seen to the heart of the mystery. They knew not just how insects develop and grow, but how *all* life grows. This was the fruit of attending God's message even when he chose to express himself through the least likely of messengers.

Think of how people furnish a home, the argument went. Anyone might make a fuss over the chairs and rugs in the living room, where visitors were sure to congregate. But if a seldom-visited back bedroom was impeccable, too, *that* spoke of genuine care and diligence. So it was, scientists reckoned, with insects—even a slapdash creator might do his best on a handful of showy creations, comets and chestnut trees and human beings. But only an infinitely patient, infinitely skilled artist would employ his full powers on the tiniest, most hidden facets of the natural world. "The Deity is as conspicuous in the structure of a Fly's paw," one French naturalist declared, "as He is in the bright Globe of the Sun itself."

Book after book set out the message. The title of one—*The Theology of Insects, or Demonstration of the Perfection of God in All That Concerns Insects*—spoke for all the others. The lowliest creature inspired hymns to Divine Wisdom. "What has been said of the lilies of the field may be justly ascribed to the Ichneumon fly . . . ," one writer exclaimed. "Solomon in all his glory was not arrayed like one of these."*

God's delicate craftsmanship demonstrated a virtuoso's talent, just as a spinning wheel in a doll house would be more impressive than one built to an ordinary scale. But the *breadth* of God's handiwork—the sheer number and variety of insects, each one perfectly constructed—made the story yet more impressive. One early scientist expressed his awe in rhyme:

---

*This same ichneumon (which was really a wasp rather than a fly) horrified Darwin and further convinced him that nature had *not* been designed by a benevolent God. The female wasp injects a caterpillar with a poison that paralyzes it; then she lays her eggs on the caterpillar's body. The wasp larvae devour the still-living caterpillar little by little (because a living host is more nutritious than a dead one), saving the heart for last so as to prolong the feast. "I cannot persuade myself that a beneficent and omnipotent God would have designedly created" these wasps, Darwin wrote, and in another letter he lamented "the clumsy, wasteful, blundering low, and horribly cruel works of nature!"

FIGURE 13.1. Artists as well as scientists marveled at the intricacies of insect design. This meticulously observed watercolor, from around 1575, is by the Flemish artist Joris Hoefnagel.

> God is greatest in ye Least of things
> And in the smallest print we gather hence
> The world may Best read his omnipotence.

Jan Swammerdam, the pious Dutch anatomist who flourished in the 1660s and '70s, was among the greatest of the scientists who took insects to be coded messages from God. "And God created great whales," the Bible tells us. Swammerdam preferred to point to the other end of the spectrum. "I offer you the Omnipotent Finger of God in the anatomy of a louse," he exclaimed, for in that tiny body "you will find miracle heaped on miracle."

In the strange process of insect development, Swammerdam saw the key to human development, too, as we shall see. In particular, he saw irrefutable proof of the Russian doll theory, which he was one of the first to describe. His discoveries would win him acclaim but never contentment. Unhappy, ambitious, guilt-racked, brilliant, Swammerdam reproached himself for his base motives. "I have striven night and day to surpass others and to raise myself above them with ingenious inventions and subtle techniques," he confessed. How could his research truly honor God when his heart was so impure?

But he worked incessantly. Hour upon hour, for days on end, he stared through his microscope at the intestines of the butterfly or the sex organs of the bee. (It was Swammerdam who found, in 1668, that the "king" bee was actually a queen. The scientific world reeled in astonishment.)*

In his quest to sort out the riddles of sex and conception, Swammerdam roamed throughout the animal kingdom. "All God's works are governed by the same rules," he declared, and therefore insight might come from any direction. Swammerdam had done pioneering work on the human ovaries and uterus, as we have seen, and later he moved far beyond humans and far beyond dissections.

Taking up a single, newly fertilized frog egg, he tried to dissect it in the hope of unveiling the secret mechanism within. Swammerdam was a talented artist and a brilliant anatomist, adept at the most demanding microsurgery, but he had taken on an impossible task. In short order, he noted forlornly, he had reduced an intricate arrangement of tissue to nothing but a broken mess, "crushed, and otherwise disturbed by my handling it." To probe the egg with razor and tweezers was akin to plucking apart a spider web while wearing mittens.

Such blundering was rare. The mark of Swammerdam's style was a delicate hand and a reverential eye. Often he contented himself with painstaking observation. He studied the mating behavior of frogs with attentiveness and a kind of fondness. "The male Frog leaps upon the

---

*Considering how odd the sex lives of birds and bees are, by human standards, "the birds and the bees" is a surprising idiom. No one is quite sure when it came to be a euphemism for sex, but the phrase is apparently not ancient. Cole Porter may have coined it, in 1928. (The original lyrics of "Let's Do It" skipped over birds and bees in favor of a bit of jaunty racism: "Chinks do it, Japs do it, Up in Lapland even little Laps do it." When tastes changed, Porter rewrote the lyrics: "Birds do it, bees do it, even educated fleas do it.")

FIGURE 13.2. Mating snails as depicted by Swammerdam. The spiraling shapes at the center are the animals' penises.

female and when seated on her back . . . he most beautifully joins his toes between one another, in the same manner as people do their fingers at prayer." The image presumably came immediately to the mind of the devout Swammerdam.

Snails, too, drew Swammerdam's sympathetic gaze. He described their curious mode of sex in a chapter of his masterpiece, *The Book of Nature*, and he decorated the title page of another book with a pair of snails in flagrante. Snail sex proceeds, appropriately, at a snail's pace, perhaps because there is a great deal to negotiate. Snails are hermaphrodites, and over the course of slow-motion sexual bouts lasting two or three hours, each partner tries to impregnate the other.

In Swammerdam's account of these slow and slimy hijinks, the *f*s in the place of *s*s rendered the text almost quaint. The snails go about their bufinefs, he explained, and then, "After all is finished, the little creature, having wantonly consumed the strength of life, becomes dull and heavy; and thence calmly retiring into its shell, rests quietly without much creeping, until the furious lust of generation gathers new strength." *Omne snail post coitum triste est.*

M ANY ANIMALS INTRIGUED SWAMMERDAM, BUT INSECTS OB-
sessed him. Since childhood, he had collected them with fanatic

energy, trapping and netting and trading and buying thousands upon thousands of prizes. His father maintained a famous cabinet of curiosities, a mini-museum in his Amsterdam home that featured such miscellaneous wonders as Roman coins and colossal snakeskins. Young Swammerdam had amassed his own cabinet, but he stuck almost entirely to insects.

This fascination—"an almost uncontrollable passion," in the words of one early biographer—persisted to the end of Swammerdam's life. The study of insects proved the ideal way for him to combine his anatomical skills with his mission of honoring God. By 1669, he had been out of medical school for two years. To his father's dismay, he devoted his time almost exclusively to collecting insects rather than patients. The old man cut his son off. *Who would throw away a lucrative career so that he could study vermin?*

Swammerdam waved aside such worldly questions. When others read of his discoveries, they would worship God with newfound fervor. He would slay his enemies not with the jawbone of an ass, like Samson, but with the genitals of a cheese mite or a stag beetle. "What atheist who considered the inexhaustible artistry of the internal organs of [these] animals would not be ashamed and dumbfounded, my Lord?" he cried.

Swammerdam delighted in the glistening appearance of the fat in the larva of the soldier fly, which gleamed like new-fallen snow. He likened the penis of a bee to a piece of crystal stemware. He probed hornet excrement and thrilled at the sight of fragments of newly devoured flies that now shone like gold. Truly God was the "Artist of all artists."

In June 1669 Cosimo de Medici, on a grand tour of Europe, visited Amsterdam. The prince and a companion stopped at the Swammerdam home and examined the renowned collection of wonders. Perhaps the most wondrous sight of all was a silkworm caterpillar, not part of the collection but a favorite prop of young Jan Swammerdam. For several years Swammerdam had delighted in dazzling scientifically minded observers with a tour de force of dissection.

On this June day Swammerdam picked up the caterpillar and asked his distinguished visitors to examine it closely. *Did they see wings? Did they see antennae? Did they detect any signs whatever of the future moth?* They did not. Swammerdam made a deft slit with his scalpel and gently peeled the caterpillar's skin away. Slowly, carefully, he pried at the soft body. There was no doubt—he had found wings, and antennae, and legs. They had not taken on their final form, but there they were!

This was a revelation. For starters, it explained the age-old mystery of caterpillar and moth—the two looked so different that it had always been assumed they were two animals, not one. The classical picture was not that a caterpillar transforms into a moth (or a butterfly) but that the caterpillar dies and then the moth rises to life from the carcass. In this devout age, when believers felt that every bit of good or bad fortune and every twig on every tree carried a divine message, this story seemed especially compelling. Just as it was "clearly evident," one Dutch writer observed, "that from dead caterpillars emerge living animals; so it is equally true and miraculous, that our dead and rotten corpses will rise from the grave."

Despite his own deep faith, Swammerdam despised such talk. To say that life could spontaneously arise from dead matter was to believe in chance and randomness. That was atheism, and it was sinful. Nothing could be further from true religion.

But for Swammerdam, and for generations of scientists to come, his silkworm demonstration did far more than resolve an ancient mystery. Biologists drew two key conclusions from his bravura performance. First, life proceeded not in sudden shifts but in a slow and gradual unfolding. That was true of insects, as Swammerdam had shown, and since God was a universal ruler, it was true for *all* creatures. God made no local laws.

"The analogy between the silkworm in its cocoon and the fetus in the womb was a compelling one," in the words of one modern historian, because creatures throughout the animal kingdom begin life as tiny, wormlike forms nestled inside protective coverings. "It was natural to suppose that the same process was occurring inside one as in the other—the emergence of visible form from subvisible form."

The second conclusion was more sweeping. Here was more proof—*compelling* proof—for the doctrine that developing organisms did not sprout new bits but simply revealed premade structures that had been there all along. "Legs, wings, and the rest are folded up, and as it were packed in a most intricate manner," Swammerdam explained, and he went on to emphasize that what he had observed in insects applied to life generally, and certainly to human beings.

The adult had been there from the start, preformed within the embryo and waiting for the proper moment to make its debut. Next came an enormous leap that Swammerdam and his peers regarded as but a small, logical step: what was true of each individual organism—that it was preformed—held for that organism's entire lineage, back to the beginning and on into the future. (Otherwise, the mystery of how a growing organism knew to take on its proper form had only been pushed back a generation.) Those ancestors and descendants had *all* been preformed and in waiting ever since the original moment when God had created life.

A few diehards rejected the whole elaborate theory, but they were on the defensive. William Harvey, the skeptics' great hero, had felt sure that, somehow, a tiny embryo acquired brand-new organs and structures as it grew—it did not simply unfurl ready-made bits that had been cunningly packed into a compact package. Harvey and his allies had studied chicken embryos carefully, dissecting them shortly after conception and then studying how they changed day by day. The whole intricate process played out according to a detailed timetable, as change followed change and one bit of tissue gave rise to another.

That view is remarkably close to what modern-day biologists believe. Today the development process is called "epigenesis," and legions of scientists strive to understand just how an organism's DNA directs its growth. But with no inkling of DNA or any idea that living bodies are made of cells, Harvey's "explanation" struck his contemporaries as vague and unconvincing. *What were these mysterious forces that shaped each bone and sinew of the growing animal? Where was the plan laid down? Who was in charge?*

Harvey's allies faced long odds because they could not explain the indisputable fact that living creatures developed and grew in an orderly way. Perhaps *some* patterns in nature could arise on their own; it did not seem necessary to imagine God cutting out individual snowflakes with scissors. But snowflakes all embody the same straightforward geometry. Hands and brains are not only infinitely varied but linked so that they work in perfect harmony. Could anything so intricate happen by itself? If new organisms really do start out as structureless blobs, one French scientist scolded, then they could grow up only "by a miracle which would surpass every other phenomenon in the world."

To seventeenth-century thinkers, any theory that did not highlight God's role opened the door to chance and randomness. *Anything* could happen. Unless embryos were ready-made, explained John Ray, one of the great naturalists of the age, there was not "the least reason why an Animal of one Species might not be formed out of the Seed of another."

EVEN PUTTING GOD TO ONE SIDE FOR A MOMENT (WHICH NO ONE at the time had any inclination to do), the picture of gradual development made no sense. Twenty centuries before Harvey, the Greek philosopher Anaxagoras had asked indignantly, "How could hair come from what is not hair and flesh from what is not flesh?"

No one had a good answer. Swammerdam and his fellow scientists added further objections of their own. The notion in the 1600s and 1700s was that the most basic units in the body were organs—heart, lungs, eyes, and so on—along with nerves, blood vessels, and various other connectors. All those structures were essential for life. If a new embryo did not have them from the get-go, how could it live and grow? (And, as Anaxagoras had asked, if it had not had them always, at least in rudimentary form, how could they suddenly appear?)

Nor was that all. If the parts that formed a living creature appeared one by one, as Harvey had proposed, which came first? What good would a heart do, without arteries and veins to carry the blood it pumped? And what use would those vessels be, without a heart to send the blood around the body?

Swammerdam had won the day for his fellow preformationists, but his triumph tasted of ashes. Pious and racked by guilt for not devoting all his time to prayer, he confessed in anguish that he had chased the false gods of honor and praise. His discoveries had let him "see, touch, and feel God on all sides," but that did not console him. "I have not loved Him purely and exclusively," Swammerdam wept, "but only incidentally and in so far as it gave me pleasure" and won the respect of his peers. In his despair he damned his scientific investigations as "foolish pursuits."

In 1675, at the age of thirty-eight, Swammerdam announced his withdrawal from science. He followed a French religious mystic to her island retreat off the coast of Denmark. There he devoted himself to "heavenly reflections," as his new mentor advised, but after nine months he returned to Amsterdam and his insect studies. He lived only five more years. Swammerdam died in 1680, felled by fever and entrapped, in the words of one scientific colleague, by "melancholic madness."

FOURTEEN

SEA OF TROUBLES

ALL THROUGH THE 1600S AND INTO THE 1700S, THE WAR between ovists and spermists dragged on. Ovists held a far stronger position but could not quite dispatch their foes. No one from either side challenged the merits of preformation theory itself. That doctrine loomed over the battlefield, as solid and formidable as a fortress.

The ovists had a variety of arguments to draw on. They invoked tradition: for ages past, everyone had known that life comes from eggs. They cited analogy: look at all the eggs in the animal kingdom. They appealed to science: the eminent Harvey had stood up for the egg, and, more recently, so had Steno, Swammerdam, and de Graaf. They enlisted common sense: the large, stable egg seemed a far safer place to encase countless generations of ever smaller, ever more fragile embryos than did a tiny, thrashing sperm cell.

All this made a strong case. What could be more reassuring than new findings that confirmed ancient beliefs? The spermists, in the meantime, found themselves reeling in the face of this all-fronts assault. They did have Leeuwenhoek on their side, and the newness and sweep of his discoveries made for a dash of glamour. But the ovists' findings fit together and strengthened one another, as streams merge to form a river. *Here is one creature with eggs, and there is another. Here is an animal with ovaries and oviducts, and here is a woman with structures that look just like them.* In utter contrast, Leeuwenhoek's discovery of tiny animals in semen came out of nowhere. It was new, which was noteworthy, but it stood apart, bewilderingly isolated from the rest of the emerging scientific picture. Leeuwenhoek had plainly found something. But what, precisely? Were these "animalcules" humans in miniature form? Or proto-humans that would transform into humans? Or some other sort of animal altogether?

On the other hand, the spermist side did have some points in its favor. For one, spermist theory put the spotlight back on males, which struck many scientists as altogether fitting. They argued happily that men dominated every creative field, whether it was painting or sculpture or poetry or architecture. Didn't it stand to reason that when it came to the most important creative act of all—shaping a new life—that, too, should be a male domain?

And the image of a squadron of lively, rambunctious sperm cells jolting the egg into action made an appealingly mechanistic picture that was completely in keeping with the intellectual mood of the age. Explanations that invoked levers and pumps and pushes and twists were in vogue. Explanations, like the ovists', that relied on mystical breaths and "seminal auras" met with sneers and smirks.

Even so, Leeuwenhoek was slow to win allies to the spermist camp. He had first seen spermatozoa back in 1677, but for decades many of his fellow scientists responded to his findings with skepticism rather than deference. In those early years, that made sense. Sperm cells were hard to see, even for Leeuwenhoek, who had both endless stamina and excellent vision. (Excellent, at least, for close-up work. Some scholars believe that Leeuwenhoek was nearsighted, which might have helped him focus on tiny specimens only a few inches from his face.)

His microscopes were not like the ones we picture today, with eyepieces and specimens mounted on glass slides. They were tiny,

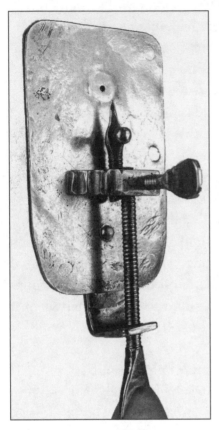

FIGURE 14.1. One of Leeuwenhoek's original microscopes.

handheld devices that consisted essentially of two metal plates that held a tiny lens between them. He made them himself, hundreds in all.* Using them was finicky, maddening work.

Leeuwenhoek fixed his specimen to the point of a needle (or, if he was looking at living creatures, put his drop of water or blood or semen inside a tiny glass tube). Then, using screws attached to the mounting pin, he manipulated the sample near to the lens. Students in high school biology classes today do things the other way around: they move the lens, and the sample sits still.

The whole apparatus was only about the size of a business card, and the lens was a miniature glass bead a fraction of an inch in diameter. (A drop of water that falls on a phone or an iPad magnifies the screen in the same way that the bead did.) Leeuwenhoek found that the smaller the lens, the sharper the view. For hour upon hour, he stared blinking and watery-eyed at his tiny samples, trying for a clear and steady view.

Even his enemies conceded that Leeuwenhoek had "the patience of an angel." To see properly he had to hold the microscope so close to his face that often his eyelashes hit against it. "On the close inspection of three or four drops," he wrote, "I may indeed expend so much labor that the sweat breaks out on me."

---

*Nearly all Leeuwenhoek's microscopes have been lost. One sold at auction in 2009 for $491,776.

Other microscopists lacked Leeuwenhoek's skill, and they lacked his microscopes, which were far better than anyone else's. Beyond granting an occasional peek by way of demonstration, he would not let others use them. Nor would he divulge his viewing techniques or his manufacturing secrets. The result was that no one else could match his achievements, but no one else could verify them, either.

Finally, around the year 1700, Leeuwenhoek gained some important allies. The spermist case took off, at least briefly. Nicolas Andry was a French doctor with a fine reputation and a much clearer prose style than Leeuwenhoek. His observations of "Spermatic Worms" confirmed Leeuwenhoek's and, in places, went beyond them. "These Worms are not found before the Age proper for Generation. They are found dead or dying in old Men, and in those who have Gonorrhea or Venereal Distempers. What must we infer from those Circumstances? Does not the thing seem to speak of itself, and tell us plainly that Man, and all other Animals, come of a Worm?"

Leeuwenhoek's most eminent ally was Gottfried Leibniz, Newton's great enemy. Leibniz was a philosopher and mathematician renowned for his astonishing intellectual range and power. One rival confessed that even the prospect of challenging Leibniz seemed pointless. Far better, he wrote, "to throw away one's books and go die peacefully in the depths of some dark corner." That judgment fit nicely with Leibniz's opinion. A man of formidable vanity, his favorite gift to couples getting married was a collection of his own sayings. But he knew everyone, as well as everything, and he wrote endlessly. This was a good man to have on your side.

Still, spermists faced a sea of troubles. Their run of good fortune lasted only from around 1700 to 1720. Otherwise, ovists dominated almost the whole of the 1700s. Strangely, much of this ascendance was not of their own doing. Though ovists happily pointed out weaknesses in the spermist case, the damage to the spermist side came mostly from unexpected directions.

From first impressions, for one thing. Eggs conjured up cheery thoughts. Sperm conjured up worms. Nicholas Andry's description of spermatozoa appeared not in a book on sex and conception but in a long volume on parasites entitled *An Account of the Breeding of Worms in Human Bodies.* By the time sperm cells made their appearance, readers had suffered through countless descriptions of a "Patient almost dead with Pain" who recovered after "a Worm came out of his right nostril, above a Span long" and "a Person, who voided Thirteen long worms alive and Woolly like a caterpillar, from his Nostrils, Ears, and Mouth."

And Andry was doing his earnest best to make the case *for* spermism. But even he admitted to shuddering a bit when he looked at sperm cells under a microscope. Dissect a testicle and take a close look at it, Andry advised, as he had done. "You shall discover in it such a hideous number of little worms that you shall hardly be able to believe your own eyes."

It was hard to imagine that so grand a subject as new life could have its origin in anything as vile as a sea of writhing worms. Man was made in the image of God, after all. From God forming a man of the dust of the ground to humankind wrapped up inside a worm was a mighty fall. It was bad enough that, in Shakespeare's phrase, we end as "food for worms." To say that we *begin* within a worm as well was too grim to contemplate. The notion seemed to make our lifetimes little more than a bizarre interlude. *Hail humankind, the once and future worm!*

For Andry and his contemporaries, microscopic life horrified in two different ways at once. Blindly twisting mini-creatures were revolting in their own right, first of all, and they never appeared on their own but always in vast numbers, which made matters even worse. To the naked eye, a dewdrop shimmering in the sunlight was an emblem of fragility and elegant design. But peek through the microscope. Thousands of living creatures jostled for space in every drop of water, in every fleck of blood, and, in particular, in every jot of semen. This was not exhilarating, a vision of life where there had been emptiness and sterility, but a glimpse of chaos and oppressive crowding where there had been pattern and order. Under the micro-scope's lens, the world was transformed from a luxuriant park to a jammed and sweltering subway car.

Making matters worse for the spermist camp, they could never quite decide what Leeuwenhoek's tiny animals *were*. Their purposeful swimming, which was the very clue that made it seem likely they had something to do with generating new life, raised all sorts of perplexing questions. Leeuwenhoek and the other spermists took for granted that these "tiny animals" were indeed animals, but what kind? If they made their first appearance in a teenage boy, where had they been for the previous dozen years? Did these strange animals mate and reproduce? Most important, exactly what did it mean that each spermatozoa con-tained a human embryo inside it? Were sperm cells animals that *became* miniature human beings? Or animals that *contained* human beings? (For the ovists, matters were a bit simpler, because everyone knew that a living creature could develop inside an egg.)

Leeuwenhoek believed he could see two different sorts of sper-matozoa, which he took to be male and female. The two sorts did not mate with one another, he suggested, but the males gave rise to male babies and the females to female babies. Precisely how that happened was, for spermists, a profoundly puzzling question. The most notorious suggestion came in 1694 from a Dutch microscopist and mathematician named Nicolaas Hartsoeker. "Each little animal actually encloses and hides an even smaller being under a tender and delicate skin," he wrote. Sex between a man and a woman serves as a kind of transport system for bringing that little animal together with an egg, which would nurture it. "Perhaps if we could see the small animal through the skin that hides it, we would see something as this figure represents."

Then followed a drawing that generations of histories and textbooks have made notorious ever since, of a big-headed person curled up inside a sperm cell, hands clutching knees as if he has just been told to brace for a crash. But Hartsoeker's cautionary "perhaps we would see" is important.

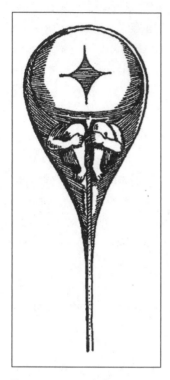

FIGURE 14.2. A figure in a sperm cell, 1694.

He did not say that he had seen this miniature figure, but only that, some day, someone might.

Skeptics delighted in imagining these miniature human beings tucked inside sperm cells. One physician entertained lecture audiences with a vision of "little men and little ladies, striking about and playing in the male semen, each of them endeavoring to get first into the ovarium and from thence into the womb, so that in time they may become fine ladies and gentlemen, princes, prime ministers, lawyers, heroes."

The spermists did their best to fend off the mockery. In 1700, a few years after Hartsoeker's suggestion, Andry made a point of emphasizing that it would be incorrect to think of "the spermatic liquid of dogs containing little dogs, that of cocks little cocks, and that of humans little children." This was not the all-out retreat it might seem. Andry and other spermists still held to the Russian doll model. Their point was merely that since no one had yet seen those dolls, no one knew just what form they might take.

But the most perplexing riddle of all, for spermists, was different. Why had God ordained a system that relied on millions upon millions of living, tiny animals to do the work of just one? The numbers, even if we put the question of Russian dolls to one side for a minute, were astounding. By Leeuwenhoek's reckoning, the number of "animalcules" in a single ejaculation was on the order of hundreds of millions.

Scientists today put the figure at about 250 million, which fits neatly with Leeuwenhoek's estimate.* Big numbers are hard to grasp,

---

*Why it takes such a fantastic number of sperm cells to fertilize one egg is still a matter of scientific debate. (For generations, medical students have learned that the reason is that, "None of them will stop to ask for directions.")

but 250 million seems *absurdly* big. *War and Peace*, for instance, contains nowhere near 250 million letters. It would take eighty copies of *War and Peace* to total 250 million letters. Think of the overkill if just one letter from that teetering stack of books could somehow convey Tolstoy's entire novel.

In the 1700s, this notion of waste was an abomination. God's designs were perfect, not ludicrously wasteful. Moreover, the notion of one winner and millions of also-rans inevitably conjured up thoughts of lotteries. Had God designed a system of conception that relied on *chance?* Even the suggestion was heresy.

The ovists pounced. From the correct observation that there are millions upon millions of sperm cells, they leapt to the incorrect conclusion that sperm cells cannot play a key role in conception. Here was still more proof that the spotlight belonged on the egg, not the sperm. The male does contribute something to conception, the ovists conceded, but that contribution must be semen rather than the sperm cells within it.

Worst of all for the spermists, worse than waste and chance, was mass slaughter! If the tiny animals were scaled-down human beings, how could it be that all but one of them was doomed? "With this doctrine," one shocked writer observed in 1698, one accused "the sovereign Ruler of having carried out an infinite number of murders or created an infinite number of useless things by forming in miniature an infinite number of men destined never to see the light of day."*

Leeuwenhoek struggled to respond. The multitude of sperm cells made sense, he argued, when you considered the size of these infinitesimal but intrepid explorers. "The womb being so large in comparison of so small a creature," he wrote. " . . . There cannot be too great a number of adventurers." He turned to his go-to example, the apple tree. It produces countless seeds, and only one or two grows into a tree. That didn't sway many listeners. An apple seed is not a human being with a soul. Nor are the extra seeds wasted; they feed birds,

---

*Three centuries later, Monty Python echoed the sentiment, this time in song: "Every sperm is sacred. / Every sperm is great. / If a sperm is wasted, / God gets quite irate."

squirrels, and mice. More important, the fate of apple seeds is not fixed. There is no reason that they can't all thrive, if conditions happen to cooperate. But if the spermists had it right, God's *plan* called for the extermination of every animalcule but one. That was a horrifying charge, and one not to be sidestepped by a debater's ploys.

Other attempts at explanation fared just as poorly. Many decades after Leeuwenhoek but still struggling with the same objection, an English doctor named James Cooke suggested that perhaps the millions of superfluous sperm cells "do not absolutely die." Instead, they might live "in an insensible or dormant state, like Swallows in Winter, lying quite still like a stopped watch." While in this hibernating state, Cooke continued, the sperm cells float randomly through the air, like dust motes, until they happen to drift "afresh into some other male Body of the proper kind." Back in the game again, they would "run a fresh chance for a lucky Conception," over and over again, forever.

This was, in fact, an old idea that had long ago been briefly considered and decisively rejected, on the grounds of sheer implausibility. Cooke's return to this played-out theory was proof of the spermists' predicament, not evidence of one bold thinker's ingenuity.

STAGGERING ALREADY, SPERMISTS FOUND THEMSELVES HIT BY ONE more wave of bad fortune. In the 1700s, Europe was caught up in a medical panic. The crisis arrived in 1712, crested in 1762, and retained its force deep into the 1800s. The issue was an age-old misdeed: masturbation. It had been condemned since biblical times, but with hugely varying degrees of fervor in different eras—a mortal sin, a minor vice, a silly indulgence, an insult to the divine plan (because it provided sexual pleasure without any hope of reproduction).

Most of the ire was directed at men. Attitudes toward their misbehavior changed so sharply partly because attitudes toward semen zigzagged through the ages. Never merely a bodily product, semen was a magical and quasi-divine concoction, if one listened to Aristotle and his countless followers, or literally demonic, if one heeded Aquinas or legions of theologians. Often, as befit a substance that was plainly important but poorly understood, semen was praised and damned simultaneously.

It was conventional doctrine throughout the Middle Ages, for instance, that demons gathered semen and shaped it into human form. Then, suitably disguised, they tempted the unwary into various forms of diabolic sex. (They went to so much trouble because they did not have bodies of their own.) All this was so well-known that it went almost without saying.* Theologians focused on logistical questions instead. Precisely how did demons obtain their semen—from masturbators? from noctural emissions? by disguising themselves as women, and seducing men? In any case, the seed spiller was abetting the devil's nefarious schemes.

But not even the church's most frenzied visions of masturbators roasting in hell had inspired a panic like the one that began in the early 1700s. The difference was that the new anti-masturbation diatribes dealt with their readers' earthly bodies, not their eternal souls, and the prospect of drooling idiocy and painful death provoked more terror than threats about burning in everlasting flames. The scare took hold around 1712, with the publication of an anonymous pamphlet called *Onania, or, The Heinous Sin of Self Pollution and All Its Frightful Consequences.*†

The thin pamphlet, which was endlessly reprinted, consisted mainly of anguished letters supposedly from victims of the heinous sin. The sufferers described in great detail the ravages of madness, starvation, and paralysis. But all was not lost. By good fortune, the author had access to several "medicines of great efficiency" that would

---

*It was well-known, too, that the children born of sex between demons and humans often had uncanny powers. Merlin, the magician who served as King Arthur's mentor, for instance, was said to have been the offspring of a nun and a demon.

†The anonymous pamphleteer coined the word "onanism," still occasionally used as a synonym for "masturbation." In the Bible Onan's father commands him to have sex with his brother's widow. But the plan went awry, we read in Genesis, and when Onan "went in unto his brother's wife, he spilled it on the ground." This sounds more like coitus interruptus than masturbation, but our pamphleteer was evidently not in the mood for a debate. Neither was God. "The thing which he did displeased the Lord, whereupon he slew him." (In more recent times, Dorothy Parker named a canary Onan, because he spilled his seed on the ground.)

FIGURE 14.3. A habitual masturbator, as depicted in an 1847 medical text.

return even the weakest to health. Contact the publisher.

Decade after decade, medical writers barraged the reading public with horror stories along similar lines. "Every seminal emission out of nature's road—I must speak plainly, gentlemen!—every act of self-pollution is . . . an earthquake," warned one London physician, "a blast, a deadly paralytic stroke." Esteemed and somber philosophers joined the baying throngs. Rousseau warned against "the most deadly habit to which a young man can be subject." Kant proclaimed masturbation more sinful than suicide.

By far the most important and influential of these nay-saying authorities was Samuel Tissot, one of Europe's most acclaimed physicians. In 1762 he produced a tome called *A Treatise on the Diseases Produced by Onanism.* This seminal work inspired genuine fear; it was as if one of our most trusted medical authorities, perhaps Dr. Spock, the author of *The Common Sense Book of Baby and Child Care*, had produced a documentary film to warn the public about zombies.

In somber but urgent tones, Tissot discussed one case history after another. One of his patients was a seventeen-year-old watchmaker. Tissot found him lying in bed almost unable to move, pale, emaciated, and "more like a corpse than a human being." The unfortunate young man had lost his memory almost completely, though he retained enough strength to acknowledge the vile habit that had brought him to this pass. "A pale bloody discharge issued from his nose; he foamed at his mouth; was affected with diarrhea and voided his feces involuntarily; there was a constant discharge of seminal fluid." Within a few more weeks, he was dead.

The danger in all such cases was the waste of semen. (Masturbation posed life-threatening dangers to women, too, but they had a bit more time to reform, Tissot explained, "the secretion which they lose being less valuable and less matured than the semen of the male.") So precious was semen, Tissot explained, that the loss of a single ounce weakened the body as much as the loss of forty ounces of blood.

At the same time, then, spermist doctrine held that nearly every bit of semen was unnecessary, and Tissot insisted that every drop of semen was precious. This contrast was not quite fair, because the waste that Leeuwenhoek and other spermists had in mind was of individual sperm cells; Tissot's waste was of drops of semen. But an epidemic of fear was no time for fine distinctions. A doctrine that waste is part of God's plan had little chance in an era that preached that waste was a physical and moral catastrophe.

So ovism won the day, and by the early 1700s spermism slunk off the field, defeated. This was, the historian Jacques Roger pointed out, a surpassingly odd development. In this era no one had ever seen a mammalian egg, and yet nearly everyone took for granted that eggs were the key to the mystery of human reproduction; nearly everyone *had* seen spermatozoa, but nearly everyone rejected sperm cells as irrelevant to the whole sex and conception riddle.

But the ovists had little time to toast their victory. From a small town south of London, called Godliman, came amazing news! An illiterate woman named Mary Toft, the wife of a clothworker, had gone into labor. Her story would provide stark proof that the ovists had celebrated too soon—sex and conception remained absolutely baffling.

It was October 1726. So astonishing was Mary's story that the king himself sent his doctor to investigate. Newspapers and pamphlets churned out countless breathless updates. All London clamored for more.

By the time things played out, Mary Toft was in prison, and Europe's scientists were in turmoil.

FIFTEEN

THE RABBIT WOMAN
OF GODLIMAN

IN THE FALL OF 1726, MARY TOFT WAS TWENTY-FOUR YEARS OLD and a mother of two small children (a third had died of smallpox). Those births had drawn scarcely any notice. But her most recent delivery brought the world running. Mary Toft had given birth not to a baby but to a rabbit! So her surgeon and midwife, a man of thirty years' experience, announced.

For us, the significance of Mary Toft's story is that it offers up the most damning proof imaginable that, as late as the 1700s, science had scarcely begun to sort out the riddles of sex and heredity. If a woman might give birth to a rabbit, then medicine and biology were truly at a loss.

After that first rabbit, Mary delivered a litter of sixteen more bunnies over the course of the next month, at a rate of about one a day. "Every Creature in town both Men & Women have been to see & feel her," wrote John Hervey, a well-known figure at the court of King George I. "All the eminent physicians, Surgeons, and Men mid-wifes are there Day and Night to watch her next production."

The learned authorities were split between those who dismissed the talk of rabbits as ludicrous nonsense and staunch believers who churned out medical pamphlets with such titles as "A Short Narrative of an Extraordinary Delivery of Rabbits Performed by Mr. John Howard, Surgeon at Guilford." Howard, the obstetrician who had reported the amazing news to England's most acclaimed scientists and to the king's doctors, had been skeptical at first. Soon he grew so certain the births were genuine that he sent off bits of the newborn rabbits to the Royal Society, preserved in formaldehyde. (None of the rabbits was born alive.)

It should not have taken sensational stories of rabbit births to highlight the message that science had no idea how parents passed on traits to their offspring. Even at this late date, the most run-of-the-mill birth still gave rise to questions that no one could answer. *Why do children so often resemble their parents? How does that work?* These were ancient puzzles, but ovists and spermists alike had neglected them in order to focus on their squabbling with one another.

Then along came Mary. Her outlandish claims served as a kind of shout that brought a clamorous room to a sudden hush. *Enough! What about family resemblances?* Mary Toft told a simple story. When she was five weeks pregnant, she'd been weeding a field when suddenly a rabbit had jumped in front of her. She ran after it but couldn't catch it. A few days later the same thing happened again. Ever since, she'd been obsessed. Too poor to buy meat, she had dreamed endlessly of rabbits. That constant musing, she said, had no doubt shaped the infant in her womb.

The hoax itself was even simpler than this cover story. Shortly before the rabbit births, Mary had miscarried, and she still showed some signs of pregnancy. Now she stuffed bits of cut-up rabbit inside her when no one was watching, and then feigned contractions. Howard, the gullible midwife, "delivered" the furry bits of flesh.

Remarkably, many people found nothing implausible in Mary's story. Everyone knew that the sights and sounds that a pregnant woman witnessed had considerable power to influence the offspring she was carrying. In the decade of the 1740s, twenty years *after* the Mary Toft case, the *Gentleman's Magazine* in London carried ninety-two such stories. One woman in Chelsea had visited the lions in the Tower of London and been "much terrified with an old lion's noise." Soon after, she gave birth to an infant with a "nose and eyes like a lion . . . claws like a lion instead of fingers, no breastbone . . . one foot longer than the other."

More than a century later, belief in this doctrine of "maternal influence" still ran strong. Joseph Merrick, the Elephant Man, wrote matter-of-factly in the 1880s that "the deformity which I am now exhibiting was caused by my mother being frightened by an Elephant; my mother was going along the street when a procession of Animals was passing by, there was a terrible crush of people to see them, and unfortunately she was pushed under the Elephant's feet, which frightened her very much; this occurring during a time of pregnancy was the cause of my deformity."*

This was a belief with ancient roots. The Bible tells of a deal Jacob made to divvy up newborn sheep and goats with his father-in-law. Animals with speckled coats went to Jacob, those with solid colors to Laban. Jacob peeled bark from tree branches to expose the white wood beneath and then placed the streaked, speckly branches near the flock. "They mated in front of the branches," we read in Genesis. "And they bore young that were streaked or speckled or spotted."

Through the millennia, pregnant women had feared the sight of hares, which were well-known to cause harelip (now known as cleft palate). They tried not to glance at the moon, for fear of giving birth to lunatics. They knew that a glimpse of a strawberry might lead to the birth of a baby with a strawberry birthmark. Worse things could happen—one doctor explained that a woman had given birth to twins joined at the head because, when she was pregnant, she had often met

---

*The true cause of the lumps and deformities that afflicted Merrick has never been settled. Modern-day doctors studying his case had come up with a diagnosis of neurofibromatosis, where the body grows (usually benign) tumors, but lately that theory has come into dispute.

with a friend and, in the course of chatting, the two women had leaned their heads close together.

Sometimes maternal influence could work in your favor.* Women knew that gazing at a tray decorated with a drawing of a healthy baby boy improved the odds that they would give birth to a boy. (Historians have never found a tray showing a girl.) In wealthy homes, with portraits on the walls, women who had become pregnant in the course of an affair tried to make things right by spending long, nervous hours staring at paintings of their husband, so that the baby would look like his father.

In one particularly desperate case, a French noblewoman named Magdeleine d'Auvermont had skipped the portrait-gazing and resorted to an even thinner tale. In 1637, in Grenoble, France, she gave birth to a baby boy. Her husband had been abroad for four years. Madame d'Auvermont found herself on trial for adultery. (Relatives on her husband's side, fearful that the baby would inherit his father's land and title, had brought the case.) Madame d'Auvermont cited the well-known power of maternal impressions—she'd had an especially vivid dream of having sex with her husband, she testified, and almost at once found herself pregnant with their child. The court ruled in her favor, the adultery charge was dismissed, and the newborn baby was officially declared the heir to the family fortune.

MARY TOFT'S SCHEME FELL APART WHEN AN ALLY WAS CAUGHT smuggling rabbit parts to her. She confessed to the hoax and was duly punished, though the authorities seemed reluctant to draw extra notice to the affair. Almost at once the public swung its attention away from the miraculous births and onto the gullible doctors. William Hogarth published an elaborate drawing called "The Wise Men of Godliman," showing astonished doctors attending a woman in labor while a swarm of rabbits scurried around the floor. Mocking the foolishness of pompous physicians proved nearly as entertaining as waiting for

---

*In our own day, there was a vogue for pregnant women to listen to Mozart, on the grounds that classical music in utero produced brighter children.

FIGURE 15.1. In this Hogarth print, a midwife tends to Mary Toft. While her newborn rabbits scurry about, one joyous onlooker exclaims, "A great birth."

word of new rabbits, and the case dragged on for a few more weeks. But bewilderment about just how inheritance worked was destined to linger for centuries.

For several reasons, heredity posed huge problems to early scientists. Armed with countless observations, many of which seemed to contradict one another, they found themselves adrift in an ocean of anecdotes. Blue-eyed parents always had blue-eyed children, for instance. So far, so good. But brown-eyed parents sometimes had brown-eyed children and sometimes did not. Why was that? Was there something special about the color blue? Something suspect about brown-eyed parents?

Parents surely passed traits to their children, but no one could imagine how that worked. A mother's broad chin might reappear in perfect miniature on her young daughter, but then another child born to the same woman would look completely different. Aristotle noted the case of a father who had been branded on the arm and had a son with the same sort of mark on *his* arm. Was there a law lurking there? Two millennia later, Darwin was still examining the same riddle. "I have been assured by three medical men of the Jewish faith," he wrote, "that circumcision, which has been practiced for so many ages, has produced no inherited effect."

Matters grew more confusing still when you followed a particular trait as it worked its way through several generations in one family. Aristotle puzzled over the case of a white woman who had gone to bed with "an Aethiop" (i.e., a black man). She gave birth to a pale-skinned daughter, but, to Aristotle's bewilderment, "the son of that daughter was an Aethiop."

If these early scientists had known that there was a science of heredity— if they had known, say, that there were strict rules that governed the color of a child's eyes—then they would quickly have seen that both mother and father played key roles in forming the new baby. But in the absence of such insight, it was easy to make up after-the-fact explanations to justify whatever beliefs you happened to hold. *That baby looks like his grandfather because the mother spent years taking care of the grandfather.*

That seemed good enough because, for thousands of years, heredity had been seen more as a collection of intriguing observations than as a mystery in search of an explanation. Our forebears thought of heredity in roughly the way that we think of trees or clouds. We know there are broad patterns—oak trees look different from maples—but no one knows precisely how the branches of a given tree will twist and turn or even how many branches there will be. We know that cumulus clouds, which feature in every child's drawing, look fat and puffy, while cirrus clouds are thin and wispy. But no one claims to know—or even *seeks* to know—just how many clouds will appear in the sky today or precisely how they will change shape as the day passes.

D OCTORS AND SCIENTISTS ACCEPTED THE MOST FAR-FETCHED stories, at least as possibilities. Eyewitness testimony commanded special attention. Odd as our forebears' credulity sounds to us—*Where was the proof?*—we can still find echoes of it in courtroom trials today, where eyewitness testimony carries far more emotional punch than technical evidence about tire tracks and bits of fiber.

One renowned doctor—this was Fortunio Liceti, a professor of medicine at the University of Padua (and a friend and colleague of Galileo)—wrote about "a scoundrel who coupled with a cow." From this dubious union, "there resulted a boy resembling a complete man in every regard, except for his sharing the cow's inclination to graze on the grass and to chew its cud."

Such lurid stories were supercharged versions of familiar tales. In books of natural history, hybrids popped up everywhere. They were not mythological figures like centaurs but real, flesh-and-blood creatures. The camelopard—in reality quite likely a giraffe—was perhaps best-known. That strange-looking beast supposedly came to be when a camel and a leopard mated.

Far more disturbing were the countless stories of human-animal offspring. From ancient times down through the 1700s, the most renowned scholars and physicians passed along such reports. They appeared not in Renaissance versions of the *National Enquirer* but in the most earnest and learned venues. The philosopher John Locke discussed the troubling case of one pig/human hybrid in his *Essay Considering Human Understanding*, one of the most admired works of its era. Locke pondered a variety of ethical dilemmas. *Would destroying the "monster" count as murder? Could the creature attend church?*

FIGURE 15.2. The original caption of this drawing (from 1573) explained, "A monster, half-man, half-swine."

One thick volume by an eminent French physician named Ambroise Paré—chief surgeon to two French kings—devoted a chapter to well-documented mix-and-match creatures.* Paré called his compendium *On Monsters and Marvels*, and it is itself a hybrid of medieval credulity and modern skepticism.

THROUGH THE AGES, EVERY QUESTION TO DO WITH HEREDITY HAD been difficult. Then the doctrine of preformation appeared, and the difficult became almost impossible. If parents don't form their children, who were created when the Earth was new, why should there be any resemblance at all between parent and child? Look more

---

*In Puritan New England a century *after* Paré, two men (in separate cases) were arrested when sows gave birth to piglets that looked suspiciously like the men in question. In New Haven in 1647, in the better documented case, the unfortunately named Thomas Hogg was confronted with a piglet that, in its owner's words, "had a fair and white skin, and head as Thomas Hogg's is." A second piglet had "a head like a child's and one eye like him, the bigger on the right side, as if God would describe the party." This was considered unimpeachable evidence, but Hogg refused to confess. By Connecticut law, that ruled out hanging. (No one had witnessed the crime.) Frustrated, the court sentenced Hogg to be "severely whipped" and thrown in jail. (In bestiality cases, both animal and human were punished; the law demanded that the animal be killed and added a perhaps unnecessary caution, "and not eaten.")

closely, and the problems grew even more formidable. If God created the complete set of Russian dolls at the beginning of time, how could he have known to give baby number 1,000,000 a pointy nose and curly hair just like her father's? What if the little girl's mother had married a man with a snub nose and hair as straight as straw?

The philosopher Immanuel Kant, whose prose typically fell somewhere in a range between dense and impenetrable, for once put the difficulty clearly. "If the woman had been with another man," he wrote, "she still would have produced the same children." (For spermists, Kant added, the same point would apply vice versa.) That left ovists and spermists caught in the same trap, neither of them able to explain how a baby could inherit traits from *both* parents.

Perhaps sensing their own vulnerability on the whole topic of heredity, both sides refrained from pushing too hard. Ovists, as we have seen, attributed nearly every puzzling observation to "maternal influence." This was a weak argument, although it helped a bit that the dolls within a doll were so astonishingly small; perhaps a tiny push *could* reshape so tiny a body, much as the flick of a finger might set a pebble flying, though it would not budge a boulder.

Spermists resorted to the ancient woman-as-field analogy, but that didn't get them far, either. The attempt was to invoke something like the winemakers' notion of *terroir*, where the same grape makes markedly different wine depending on particular conditions of soil, rainfall, and sunlight. "This is so poor," one writer scoffed in 1707, that you might as well argue that "an Orange-Tree transplanted from Sevil to England would bear Apples."

Hybrid births, in which the parents are from different species altogether, posed a particularly daunting challenge for ovists and spermists alike. Take mules, which are the offspring of male donkeys and female horses. Mules have been familiar since ancient times. But according to preformationist doctrine, donkeys should give birth to an endless line of donkeys, and horses to horses. How could it be, then, that one day a horse gave birth to a different sort of animal?

The everyday explanation was perfectly simple: someone decided, one day, to breed a donkey and a horse. But preformationists had to tie themselves in knots to explain where this long-eared interloper had come from. Had God foreseen at the beginning of time just which Russian doll would be not a horse but a mule?

I N THE ABSENCE OF ANY BETTER THEORY THAN PREFORMATION, ovists and spermists did their best to ignore uncomfortable questions. Sometimes this strategy cost them. Leeuwenhoek, for instance, had veered near one of Gregor Mendel's key findings centuries ahead of the father of genetics. Then he turned his back on his own insight.

In a letter to the Royal Society in 1683, Leeuwenhoek had mentioned some rabbit breeders he had met. He had peppered them with questions. Often, he learned, they mated wild rabbits with domestic ones. The wild rabbits were small, gray males, the domestic rabbits large, white females. All the offspring were small and gray, like their fathers.

Leeuwenhoek pursued the matter further. What happened, he asked the breeders, if the females were *not* white? Again, so long as the father was gray, all the offspring were gray, no matter whether the mother was black-and-white or solid black. "Indeed," Leeuwenhoek exclaimed, "it has never been seen that any such young rabbit had a single white hair or any other hair than gray."

What could it mean? In line with his spermist views, Leeuwenhoek concluded triumphantly that the moral was plain: only the male is important. But if Leeuwenhoek had gone on to perform a follow-up experiment—this time mating white, domestic males with wild, gray females—all the offspring would again have been gray, like their mothers. Leeuwenhoek would have been confused, but if he had concluded that gray always prevailed he would have been the first to discover a dominant genetic trait.

Leeuwenhoek was done in by ideological blinders, but his mistake was a natural one. We are all quick to settle for explanations that confirm our assumptions. All the more so when we can slide new facts neatly into place inside an existing, ready-made theory. Leeuwenhoek was too far ahead of his time. When he pondered the riddle of gray and white rabbits, he was musing about genetics two hundred years before the word even came into existence.

A GAIN AND AGAIN IN THE SEX AND BABIES MYSTERY, WE CAN SEE our forebears on the brink of grasping a clue that required concepts and a vocabulary they had not formulated. The riddle of heredity provides a tantalizing example, and not just for Leeuwenhoek. William Harvey, who lived before the preformation era, almost sorted it out. Trying to account for family resemblances, he wrote that "there is no part of the future offspring actually in being, but all parts are indeed present in it potentially."

This sounds like mumbo jumbo, but Harvey was stumbling to find words to convey a genuine insight. *Look at a woman's freckled nose,* Harvey had said, in effect. *Years from now, she might have a child with an identical freckled nose of his own. That has to mean that somewhere there is a ghostly realm where that freckled nose lies waiting to make its appearance.* The problem was that Harvey could not bring into focus what he had dimly glimpsed. Nor could anyone else in his era.

Harvey did not believe in a simple physical explanation of inheritance; he rejected the ancient suggestion that a baby inherited his mother's strong chin or his father's big ears from miniature chin and ear particles that somehow joined together as the embryo grew. But the baby *did* have his mother's chin. How could that be? With the idea of a genetic code completely out of reach, Harvey could only gesture toward an answer.

We can now describe, with hindsight's aid, what he had struggled to make out. Parents pass to their offspring not building blocks but instructions. But that insight belonged to a far-off future. In the early decades of the 1700s, scientists trying to solve the sex and babies mystery found themselves snarling in frustration. Their theory of the case relied on preformation, the doctrine of Russian dolls. That doctrine seemed unassailable. But, as we have seen, preformation made the simplest facts of heredity impossible to explain.

The impasse lasted until 1740. Then, one summer morning in Holland, a young man took two boys for a walk.

SIXTEEN

"ALL IN PIECES,
ALL COHERENCE GONE"

T HE BOYS HAD SPENT THE MORNING HAPPILY SPLASHING THROUGH the ponds on their parents' grand estate, scooping up prizes in glass jars to examine later. Without a magnifying glass, their finds didn't look like much. Green flecks, mostly, that floated in their jars.

Their tutor, a young Swiss naturalist named Abraham Trembley, believed that ponds and meadows made as good a classroom as any indoor space. Now Trembley and his two charges, one six years old and the other just three, gazed intently at the debris they had scooped up. Trembley peered through a magnifying glass. Was this quarter-inch-long green tube a plant—that was the boys' vote—or were those wavy, slow-moving tentacles not branches but arms? *Plant or animal?*

Trembley set to work to find out. The experiments he carried out, for two little boys and with hardly more equipment than a pair of scissors, would stun him and turn the world of science upside down. Trembley shot to fame. Today he is all but forgotten, and nobody pays attention to fresh-water polyps. In the 1700s, his discoveries were considered the most important of the age.

His first observations gave little hint of the revelations that lay ahead. Trembley saw early on that his pond creatures could crawl along the glass walls of their jars and shorten and lengthen their bodies. Then, one day when a water flea swam by, Trembley saw the polyp shoot out an arm, grab the flea, and stuff it into its "head." *Okay, then—these were animals, not plants.* That seemed noteworthy, perhaps, though scarcely earthshaking. To seal the deal with one final, simple experiment, Trembley took scissors in hand and cut one of the quarter-inch long organisms in two. If the polyp was a plant, he reasoned, then there was a chance it would survive; some plants grow from cuttings. But if it was an animal, splitting it would surely kill it.

Trembley watched and waited. After a week, both head and tail were thriving. (The head was the end with tentacles.) After two weeks, the head had grown into a full, new creature. So had the tail! One creature had become two. Trembley repeated the experiment. Same result. He tried again, now making several cuts. He watched in disbelief as each tiny snippet cut from the original creature grew into a full-fledged, independent animal, indistinguishable from its "parent."

In an era when living animals, especially small ones, were regarded as machines, this was unfathomable. It would be as if, in our day, someone took a blowtorch to a car, cut it into random pieces, and then looked on as each piece—this one containing a mutilated chunk of engine, that one a bit of door and a side mirror—regrew into a complete car and zoomed off down the street.

The first report on Trembley's "little machines," from 1741, conveyed the scientific world's astonishment. "From each portion of an animal cut in two, three, four, ten, twenty, thirty, forty parts and, so to speak, chopped up, just as many complete animals are reborn, similar to the first." So far as anyone could tell, the process could go on forever. "The story of the Phoenix who is reborn from his ashes, as fabulous as it might be," declared France's normally staid Academy of Sciences, "offers nothing more marvelous."

Trembley sent glass jars with hydra inside to all of Europe's leading scientists. (He named his creatures for the nine-headed beast that

FIGURE 16.1. Hercules battling the hydra.

Hercules had managed to kill even though two new heads grew whenever he cut off one.) Skeptical but fascinated, naturalists launched into a frenzy of experiments. But whether you cut the creature in pieces or turned it inside out or put one hydra inside another, nothing fazed this bizarre animal. "These are Truths," wrote one bowled-over English scientist, "the Belief whereof would have been looked upon some Years ago as only fit for *Bedlam*."

More bewilderingly still, the creature performed its miracles even though it seemed not to have any working parts. Even under a microscope's lens, one scientist complained in frustration, hydra appeared to be nothing but a stomach. Trembley's first thought had been that the growth of a new hydra from a mere scrap might be a souped-up version of some more familiar form of regeneration, like a crab's growing a new claw. But no, he quickly conceded, that would not do. A new claw was remarkable, but a complete new organism was uncanny.

Everyone, not just scientists, wanted in on the excitement. Sales of microscopes soared. Salons buzzed with chatter. Writers scurried to explain what it all meant or to lampoon the intellectuals who sat entranced, staring at magnified drops of water. "If 'tis cut in two, it is not dead; / Its head shoots out a tail, its tail a head," wrote one English satirist, in a mock epic. "Cut off any part that you desire, / That part extends and makes itself entire."

WHY SUCH ASTONISHMENT? NOT SIMPLY BECAUSE THESE ANI- mals acted like no animals ever had. More than that, these tiny creatures blasted the firmest tenets of eighteenth-century biology to smithereens. First, the most fundamental law of nature had suddenly been violated. *Here is new life arising without a hint of sex or mating.* Second, biology's reigning theory, the doctrine that new life came in the form of dolls within dolls, found itself shoved aside and completely beside the point. *Does anyone seriously maintain that every single piece of cut-up hydra, no matter where it came from, contained a mini-hydra within it?* Third, the all-but-universal assumption that living creatures were ingenious machines no longer made sense. *Who has ever heard of machines like these?*

Underlying those distinct challenges was a deeper assault on established views, one almost too unsettling to spell out. If life could shape itself from random scraps of matter, as now seemed undeniable, how did God fit into the picture? Where in these scenes of burgeoning life was the Creator's shaping hand?

And still more trouble was coming. At virtually the same time that Trembley was slicing up hydra, another Swiss naturalist was staring, bewildered, at aphids. These, too, were seemingly insignificant little creatures destined to undermine the foundations of seventeenth-century science.

Charles Bonnet happened to be Trembley's nephew, though he was only ten years younger. The two men had a great deal in common. Both were devout, obsessed with the natural world and with insects most of all, and in constant correspondence with each other and with a host of other scientists. It was one of the most eminent of these correspondents, a French naturalist named René Réaumur, who had goaded Bonnet into taking on a mystery that he himself had failed to crack.

Aphids are small, ordinary-looking insects whose sex lives could not be further from ordinary. Réaumur was one of *the* great authorities on insects, renowned for the breadth of his knowledge and the reliability of his observations. For several years he had been at work on what would become a six-volume *Natural History of Insects*. But in all his research, he noted, he had never seen aphids mate, and he had never even seen a male aphid. Yet aphids abounded. How could that be?

Bonnet set out to find the answer. His strategy could hardly have been simpler. On May 20, 1740, he took a single newborn aphid—female, of course—and put it on a branch cut from a bush. Then he put the branch inside a glass jar and sat down to watch. Alone in her glass prison, Bonnet's aphid went on her buggie way. Bonnet kept watch "day by day and hour by hour," he wrote, with a magnifying glass always at his eye. No one tampered with the jar. No hidden aphids sneaked their way out of the branch or crept under the glass. Still Bonnet watched.

On June 1, after eleven days in solitary confinement, the aphid gave birth. Bonnet kept watching. (He would stare through his microscope so intently, in such a variety of experiments, that by the age of twenty he had ruined his eyes. He never recovered his vision.) Over the next three weeks, more new aphids appeared. By June 24, the tally had reached ninety-five. There had been no males and no mating, but where there had originally been one lone aphid, now there were nearly one hundred.

In July 1740 Réaumur stood in front of the French Academy of Sciences and read aloud a letter from Bonnet describing his work. Bonnet's reputation was made. The young naturalist, just turned twenty, had witnessed a virgin birth.

For scientists, this was shocking, especially since the hydra news and the aphid news arrived at nearly the same moment. The point was not that these tiny creatures were important in themselves; the point was that they had upended time-honored laws of nature. These were impossible creatures, and yet there they were, in countless ordinary ponds and bushes, carrying on in their paradoxical ways. (The modern-day counterpart would be something on the lines of the discovery of a butterfly that lived forever.) Everyone knew that God had decreed fixed and eternal rules that governed all life on earth. With Trembley's indestructible creatures and Bonnet's virgin births, God's laws had been not dodged but shredded.

T HE FIRST RESPONSE WAS TO TRY TO EXPLAIN AWAY THESE DIS coveries. Perhaps, once the initial shock had passed, they were not so strange, after all? Someone now recalled that decades before, back in 1696, Leeuwenhoek and another Dutch scientist named Steven Blanckaert had both written about aphids. Leeuwenhoek had described them carefully, although in an odd context that might have distracted his readers.

At the time, Leeuwenhoek had been preoccupied with the mystery of where spermatozoa, which he regarded as animals, came from. In the course of another project altogether—Leeuwenhoek was trying to sort out why the leaves on the cherry trees in his garden had curled up and died—he looked closely at aphids he had found on his trees and currant bushes. Here, where he had not even been looking for it, was the key to the spermatozoa riddle! Aphids began as lone, tiny dots and grew into swarms of full-fledged creatures, with no mating anywhere along the way. *Perhaps spermatozoa did the same!*

Spermatozoa, Leeuwenhoek proposed, arose out of some mysterious "essential stuff" in the testicles. Even to Leeuwenhoek, that seemed dismayingly vague. Still, this was a remarkable display of intellectual agility. To rescue his doctrine that, when it came to sex, only males counted, Leeuwenhoek had devised a theory in which males, and mating, played no role at all.

More study of his aphids revealed further surprises. Leeuwenhoek dissected an aphid and discovered within it a host of miniature, unborn aphids! He had expected to find eggs within the parent's body, but to his astonishment he found instead "animals the shape of which resembled that of their Father or Mother as closely as two drops of water resemble each other." Within a single parent, he found *seventy* offspring.

And within those offspring, he detected even tinier micro-miniature aphids. This startling finding was absolutely correct; we now know that aphids do carry on exactly this way. Even Leeuwenhoek found it bizarre—"I was at my wits' end to fathom this secret of generation," he confessed—but it did seem to provide a fine example of preformation, so perhaps it made sense after all.*

IN THE 1740S, WITH THE SCIENTIFIC WORLD FLUMMOXED BY HY-dras and aphids, the mysteries of regeneration and reproduction were up for grabs. Everyone wanted in on the game. From across Europe came an orgy of cracking and breaking and splitting seldom seen outside a Baltimore crab house. Bonnet guillotined snails and found that some kinds could grow entire new heads. He chopped up worms into dozens of pieces and found that, eventually, dozens of complete, intact worms wriggled away.

Starfish, crayfish, and salamanders joined the throngs of animals sacrificing their limbs to scalpel-wielding naturalists. Straightforward experiments gave way to clever but grim variations—*What would happen if you amputated a lizard's foot and it grew back, and then you amputated that new-grown foot?* The answer, it turned out, was that the poor, tormented lizard once again grew a new foot.

For scientists in the 1700s, who took for granted that humans stood atop the peak of the pyramid of creation, these were bewildering observations. How could God have endowed crawling, scuttling creatures like worms and crabs with powers that human beings lacked? A human who lost his head was lost indeed; a beheaded snail shrugged his puny shoulders and carried on.

---

*Nearly three hundred years later, biologists were still wrestling with the aphid mystery. Aphids turn out to be startlingly versatile beasts, capable of reproducing both sexually and asexually. When times are good, all aphids are female, and they churn out identical copies of themselves as quickly as possible, without mating. When times are bad, the mothers give birth to both males and females. They reproduce by mating, as if figuring that sex will yield all sorts of offspring, some of whom may thrive in the unpredictable new world.

Nearly as puzzling, these stories of magical regenerative powers seemed to convey one message, while the aphid experiments conveyed a contrary one. On the one hand, the stories of regeneration undermined the theory of preformation. If any old piece of worm could give rise to a full worm, did that mean that worms in nested versions sat hidden everywhere inside a worm's body? No one thought so. The lizard experiments posed a similar challenge. Did lizard *feet* come packed inside one another, at the ready for every devilish contingency, and with backups to the backups? Even the staunchest believers in preformation felt uneasy.

On the other hand, the discovery of aphids within aphids seemed to speak *in favor* of preformation. Still, a lone species of insect pest didn't carry much clout. The skeptics outshouted the preformationists, confronting them with embarrassing challenges, mostly to do with heredity. Similar objections had been raised before, but the preformationists had managed to deflect them. Now the attackers moved in again. One eminent French scientist, Pierre Louis Moreau de Maupertuis, took the lead. Maupertuis had a taste for combat and a gift for posing vexing questions. Worse, he spoke from a position of authority.

Like many scientists in the 1700s, Maupertuis roamed from field to field. An astronomer and a physicist before he turned his attention to biology, Maupertuis had been a celebrity since the 1730s, when he had led a French expedition to the Arctic to sort out a controversy about the shape of the Earth. It turned out to be a squashed globe, pushed in a bit at the poles and bulging at the equator. This was what Maupertuis (and Isaac Newton) had predicted, and soon no literary evening or elegant dinner or royal ball was complete without Maupertuis.

Witty, vain, elegant, seductive in print and in person, Maupertuis spoke as if he had done the squashing himself. He courted controversy. "He who causes himself to be often spoken of is always discussed," he advised a friend, "and that is everything." Heeding his own advice, Maupertuis had shifted his focus from geography and geometry to sex and heredity.

In about 1750, he found a family in Germany in which, over the course of four generations, babies had been born with six fingers on each hand and six toes on each foot. For preformationists, that was an enormous riddle. When God created his Russian dolls at the beginning of time, why in the world would he have made some of them with extra fingers and toes? And yet Jacob Ruhe *had* been born with extra digits, and so had his mother, and her mother before her, as well as three of Jacob's eight siblings. And when Jacob married (a woman with standard-issue hands), two of their six children had extra digits.

Maupertuis tried to calculate the odds that a string of such unusual births could occur by chance. He surveyed the population as best he could and decided that, at most, one person in twenty thousand was six-fingered. A bit of multiplication led him to conclude that the odds were trillions to one against the possibility that all those Ruhes just happened to have six fingers. If that was coincidence, then "the best proven things in physics" were coincidence.

The story of the Ruhes cried out for some explanation involving traits passed from parent to child, Maupertuis insisted, but preformation ruled out such accounts. Odder still, the extra-finger trait seemed to follow the father's line in some generations of the Ruhe family tree and the mother's line in others. Neither ovism nor spermism permitted such things. Maupertuis gloated, and his preformationist rivals fumed.

Still not satisfied, Maupertuis and other skeptics mounted yet another attack on the preformationists. This time they focused on what the eighteenth century insisted on calling "monsters." How did the preformationists explain birth defects? Could God really have stocked his Russian dolls with twisted, malformed infants, a nightmare nursery of blind and hunchbacked babies? One indignant ally of Maupertuis fumed at the idea that God had designed "monstrous eggs" at the Creation.

The preformationists fell back under these assaults, but they were hardly routed. The argument about "monsters," for instance, was easy enough to counter. Did Maupertuis presume to instruct God on how to manage creation? God's ways were not for men and women to judge. Everyone had always known that fate was cruel and the world a vale of suffering. What was new there? And surely God did not avert his eyes from pain. Hell, after all, was a place universally believed in, and countless souls endured endless torment there.

But the intellectual tide had turned. Arguments that had once been cited as strengthening the case for preformation were now invoked against it. The most important was the infinite sequence of dolls within dolls that lay within the egg or sperm. Originally, that infinite chain had testified to God's infinite power. Now the same observation was cited as highlighting preformation's absurdity. The Count de Buffon, one of the great mathematicians of the 1700s, took the trouble to calculate just how tiny those Russian dolls were. By his reckoning, they grew tiny impossibly fast. Even if you started with a Russian doll as big as the entire *universe*, within a mere six generations—by the sixth doll inside a doll—you would need a microscope to find the smallest doll. And six generations was hardly any time at all; *hundreds* of generations had passed since Adam and Eve.

THE ASSAULT ON PREFORMATION WAS A TEAM EFFORT, WITH Trembley (and his hydra) recruiting Bonnet (with his worms and snails), who enlisted Maupertuis (with his six-fingered families), who tapped Buffon (with his calculations about infinitesimally small Russian dolls). Preformationists reeled under those attacks. But even though they had no good answers to most of these challenges, believers in preformation had come closer to the truth than they knew. They had sensed correctly that much of a person's destiny—whether she will have curly hair or dark eyes or straight teeth—is imprinted on her from her earliest days.

Their problem was that they could not imagine how that information could be conveyed. How could they, without any useful analogies to draw on? Harvey had his pump, Newton had his clockwork, but biologists in the 1600s and 1700s had no technology they could look at to spur their imagination. What they lacked was any example of a machine that could follow instructions written in code.

Scientists by this time knew how to make all sorts of sophisticated mechanical devices. Clocks and watches could be started up and set running, almost as if they were alive, and engineers had outfitted palaces and country estates with arrays of fountains that spurted water high into the air in carefully choreographed sequences. But those machines always did precisely the same thing, over and over again (that was, after all, the point of a time-keeping machine).

In the winter of 1738, all Paris had lined up to see a mechanical marvel that represented the height of the inventor's art. This was a metal duck, complete with copper feathers, that could stretch its neck toward a visitor, take a kernel of corn from his hand, swallow it, and— here was the great, spectator-pleasing touch—"discharge it, digested, by the usual Passage." Two other automatons, one a drummer whose mechanical hands tapped out a rhythm with his drumsticks and the other a flutist who tweeted away, competed for notice.

All three figures enthralled the crowds. (The flutist contained an "infinity of wires and steel chains . . . ," one admirer wrote, "[which] form the movements of the fingers in the same way as in living man, by the dilation and contraction of the muscles.") But the defecating duck was the star of the show, even though his inventor admitted that matters were not quite what they appeared. The duck did not really digest his food and excrete it, Jacques de Vaucanson admitted, but simply ground it up and left it sitting at the base of its mouth tube; at each meal, the duck's tail-end had to be loaded separately with soon-to-be-excreted pellets.

Vaucanson's mechanical duck made a fitting emblem for the state of biology in this era. On the one hand, its wing-waving, tail-flapping, bill-dipping performance looked uncannily ducklike. An ingenious designer, it was now plain, could build a machine that would charm the grouchiest spectator. On the other hand, all that ingenuity only highlighted the gulf between the most sophisticated cogs-and-gears machine and the most ordinary living duck. Vaucanson's duck was a marvel, but it had nothing to do with *life*. A leap in the air was a feat, but it was not progress toward a voyage to the moon.

So, for eighteenth-century scientists trying to understand how a baby comes to resemble its parents, the problem was not that they lacked mechanical know-how. Nor was the problem a failure to grasp the importance of codes. The idea that a string of mysterious symbols could carry a message was an old one. The message might be *Attack at dawn!,* but everyday examples abounded, too. The earliest musical notation dates from the Middle Ages, for example, and notes drawn on a staff are a kind of code telling a musician what to sing. Written language, for that matter, conveys meaning through a sequence of cryptic lines and curves drawn on a page. Learning to read *is* code breaking.

The problem for scientists grappling with the riddles of heredity was that they could not come up with an analogy that might have spurred their imagination: they had never seen a machine that could follow instructions and proceed down one path rather than another. If they had seen such a fantastic machine, perhaps they would have wondered if a human being, too, could grow and develop by employing the same strategy.

Programmable machines would come along in the next century, in the early 1800s. The player piano was one. It could plink out any tune at all, from a lullaby to a Bach minuet, depending on the roll that an operator gave it. And automated looms, controlled by punchcards that specified various designs, would weave carpets in an endless variety.

Those new machines would arrive a century before the first modern computers, but player pianos and similar inventions might conceivably have appeared even earlier than they did. In many elegant homes in the 1700s, visitors looked on as their proud hosts displayed an amusing contraption called a music box. Only a narrow gap separated that music-producing machine from a player piano, but nobody made the jump. (A music box typically played only a single tune, which depended on the arrangement of pins on its revolving wheel, but there was nothing to rule out a system of easy-to-install, interchangeable pinwheels.)

And it was as early as 1679 that Gottfried Leibniz, one of the most far-seeing geniuses the world has ever known, imagined the computer, or something close to it. But at that point even electricity was still completely mysterious—Leibniz's computer was a sort of colossal pin-ball machine, with marbles rolling down ramps—and he never built one. No one, in fact, built an instruction-following machine in the 1700s.

Had anyone done so, the scientists trying to make sense of hered-ity might have had an easier time. *How does a baby "know" to grow straight, black hair?* The preformationists had sensed, correctly, that in some sense the black hair was foreordained. If computers had been an everyday fact of life, someone almost surely would have said, *Hidden somewhere inside that tiny lump of tissue is a program specifying black hair.* Instead, they saw no alternative to the theory that the black hair itself had existed all along, in miniature form.

It is hard not to feel sorry for our baffled scientist-detectives. Un-able to imagine a technology that would not appear for decades to come, they found themselves in the predicament of a time-traveling Sherlock Holmes trying to solve a twenty-first-century murder. *"Now, Watson, let us examine the facts. We know that a man walking to his place of work in a great city carried on a conversation with another man in that city, without raising his voice, at a distance of some ten miles. The testimony on this point is incontrovertible. And, yet, it is impossible."*

This kind of forgivable blind spot features often in the sex-and-babies story. Often in science, the problem is the other way around: rather than flail about because they cannot imagine a model that accounts for the facts, scientists take the reigning technology of their day and apply it willy-nilly. In ancient Greece, the heart was a furnace. By the 1600s, it had become a pump. At the end of the nineteenth century, the brain (or the mind) was a steam engine in which the pressure of repressed memories and dark desires could lead to catastrophic explosion. In the early twentieth century, the brain was a telephone switchboard staffed by teams of operators manipulating a cat's cradle of cords and plugs. A few decades later, the brain was a clicking, whirring computer.

The cliché is right in insisting that to the person whose only tool is a hammer, the world looks like a nail. But it is also true that to a person without a metaphor, the world looks like a blur.

SEVENTEEN

THE CATHEDRAL
THAT BUILT ITSELF

F RUSTRATED BY THE QUESTION OF HEREDITY, SCIENTISTS IN THE mid-1700s turned to a related mystery. This challenge was so basic that it had been overlooked, as we overlook many things we take for granted. *Forget about black hair and blue eyes and other details for a minute, scientists now asked. What about this? How does a baby know how to grow at all? How does a helpless infant "know" to become a toddler and then a teen?*

The question was both unavoidable and unapproachable. If preformation was not the answer, where did you begin? Perhaps there was some vital force within the body that directed its growth, as had sometimes been suggested. That had a certain ring to it, but it was too vague to help much. But consider for a minute what a proper explanation would require. One of the leading thinkers of the age, the Swiss anatomist Albrecht von Haller, wrote in 1752 that he could not imagine a "force that would be sufficiently wise to join together . . . millions and millions of vessels, nerves, fibers, and bones."

A novelist and a poet as well as a scientist, Haller had talent and ambition to spare. Still, he acknowledged that when it came to the mystery of how life organizes itself, he could do little more than confess his bewilderment. Without a "building master" to supervise the assembly of the microscopic bits and pieces of the developing embryo, Haller wrote, gruesome accidents would be universal. "An eye might stick to a knee, or an ear to a forehead."

Preformation's great appeal had always been that it offered a way to deal with these objections, or at least to dodge them. How an embryo transformed into a baby was not a mystery because there *was* no transformation; there was only growth and unfolding. Just as important, preformation was a mechanistic theory, one that depicted the world as a place of pushes and pulls rather than of forces, auras, and emanations. In the early 1700s, to think in this cogs-and-wheels way marked a scientist as a sensible problem-solver and not a wooly-minded dreamer. But as preformation came under attack in the middle decades of the 1700s, the clockwork picture came to seem narrow and misguided, at least when it came to the living world.

And especially when it came to sex. Some scientists, even a few who had been the most enthusiastic about mechanistic theories of the cosmos, had been sounding the alarm for decades. Bernard de Fontenelle was a French writer famous in the 1600s and 1700s for his popular accounts of science. In his best-known book he had proclaimed flat out that "the universe is but a watch on a larger scale." Fontenelle was a worldly, sociable man who lived to within a month of his one hundredth birthday and remained a fixture at every literary salon and elegant dance until the end ("Ah madame," he sighed, on meeting one famous beauty, "if only I were eighty again!"). But despite his devotion to clockworks and to romance both, Fontenelle made a point of emphasizing that the two did not belong together.

As early as 1683, at the height of the Scientific Revolution, Fontenelle had mocked his fellow mechanists for overreaching. They had tried to import their clockwork explanations from physics into biology. Fontenelle would have none of it. "You say that Beasts are Machines, just like Watches?" he challenged. But everyone knew what happened when you put a "Male Dog Machine next to a Female

Dog Machine." Soon you had little Puppy Machines. "Whereas two Watches"—the machines that represented human engineering at its most sophisticated—"might be together all their Life-time, without ever producing a third Watch."

Such objections had not risen to the status of deal breakers early on, because clockwork explanations had been so much in vogue. That faith stemmed from physicists' success in explaining the solar system. Then Robert Hooke and Antony van Leeuwenhoek had come along and inspired hope that the microscope would do for biology what the telescope had for astronomy. That was especially easy to believe since the two inventions were so closely related. (Galileo made his observations of the microworld essentially by looking through the wrong end of his telescope.)

The pioneering generation of anatomists had shown that life's secrets were written in minute script. The microscope seemed like the perfect tool for deciphering what one scientist called the "mystical letters" in those God-dictated texts. And anatomy was only part of the story. *All* the tiny building blocks that formed the world could now be studied facet by facet. One early scientist thought that in time microscopes might reveal "the Solar Atoms of light" and "the springy particles of air."

But one of the major scientific trends of the age was the collapse of these high hopes. No matter how diligently they worked, biologists in the 1700s found that life did *not* come into sharper and sharper focus but grew ever dimmer and stranger and more baffling.

This was deeply disillusioning. The microscope's all-seeing eye, Hooke had forecast in *Micrographia*, would confirm the mechanistic picture of the world. In the past, he noted with disdain, thinkers had indulged themselves in idle guesswork and dreamy speculation. No more. With science's new tools, empty chatter would "give place to solid Histories, Experiments, and Works."

Hooke had a firm idea of what those experiments would show. We might "discern all the secret workings of Nature," he wrote eagerly in 1665, and we would see that the whole intricate system was "managed by Wheels, and Engines, and Springs [akin to those] devised by human Wit." Even memories might have an actual, physical shape. Hooke suggested they might look something like curled-up snakes sleeping in a cave, a "continued Chain of Ideas coiled in the Repository of the Brain."

Those hopes were soon dashed. First, technical problems proved maddeningly obstinate. You could not simply put your eye to a microscope and look around at the micro-landscape, like a sailor scanning the sea through a spyglass. Under the microscope's lens, colors faded, and opaque objects grew transparent. "It is exceedingly difficult in some Objects, to distinguish between a *prominency* and a *depression*," Hooke lamented, or between a reflection and a surface or a shadow and a stain.

The telescope had been easier to understand and easier to trust. You could look at a church steeple half a mile off through a telescope. Then you could walk closer and confirm that what you had seen matched reality. And when you turned that telescope to the heavens, the picture continued to make sense. The most exotic objects looked reassuringly like those we knew already—the moon had mountains, like the Earth; Jupiter had moons, like our moon; and Venus had phases, like the moon. But look through a microscope, and what you saw was baffling and unexpected.

Or hideous. We have already met scientists who cringed at the sight of wormlike sperm cells. Many other observers were just as squeamish.* Jonathan Swift, who was fascinated and repelled by science in equal measure, devoted long passages of *Gulliver's Travels* to the creepiness of a world looked at too closely. At one point in the novel, which appeared in 1726, Gulliver lifted a tiny man to his cheek for a look around. From ground level, Gulliver's skin had appeared "fair and smooth." But now the Lilliputian recoiled in horror. "He said he could

---

*Even the hard-to-faze Leeuwenhoek once confessed that he should not have moved directly, one evening, from staring at dissected oysters under the microscope to devouring a plateful for his dinner. Somehow, he noted queasily, his feast left him feeling "not as much pleasure as I should have done."

discover great holes in my skin; that the stumps of my beard were ten times stronger than the bristles of a boar, and my complexion made up of several colors altogether disagreeable."

Swift's judgment was mild in comparison with many of his peers. "If our eyesight were enlarged," one horrified writer declared in 1727, a year after *Gulliver's Travels*, "we should appear to be the most amazing Spectacle in the whole World: there should we see an infinite Number of Worms swimming in the Blood, and sallying from the Heart through the Arteries, and returning back by the Veins." Everywhere you looked, whether in eyes or nose or ears, you would find countless living, burrowing animals. "We should see not only the Brain full of them, but the Flesh abounding with them, and the very Bones perforated by them; and Thousands every Moment crawling through the Pores of the skin."

The telescope enthralled; the microscope appalled. That had not always been so. From the telescope's earliest days, the prospect of stars stretching to infinity, stars beyond counting, had inspired awed thoughts of a mighty Creator. So had the micro-wonders seen through the microscope, at least for pioneering investigators like Swammerdam.

But even Swammerdam, perhaps the most pious of all the great microscopists, found himself aghast nearly as often as he was entranced. The natural world revealed through the microscope was not only shimmering butterfly wings but swooping death and cruel, devouring jaws. "How then can we avoid crying out, 'O God of miracles! How wonderful are all thy works!'" Swammerdam exclaimed, only to lament, a few words later, "All nature is over-run, and covered with a kind of leprosy [that] weighs down our senses and disturbs our reason."

Such qualms were far from universal. But as time passed, more scientists lost faith. They continued to find new microstructures, many of them dazzlingly complex, but they could not fathom why God had indulged himself in such showmanship. Once those flourishes and curlicues had inspired hymns of praise. Now the notion of endless intricate details in a mite's intestines stirred dismay and puzzlement. God seemed less a hurler of lightning bolts than the sort of mad eccentric who devoted his energies to inscribing verses on a grain of rice.

In physics the guiding principle, at least since Newton, had been that the simplest and most elegant solution to a problem was almost certainly the correct one. This doctrine was enshrined with a name, Ockham's razor, as a reminder of the virtue of cutting theories down to their sparest form. The telescope had only strengthened that belief. Nature was mathematical, precise, and austere. But then came the microscope, and nature seemed messy, exuberant, and overflowing.

To this day, biologists new to the subject find themselves taken aback by nature's tolerance for make-do solutions. (The eye is wired backward, for instance, so that we have a blind spot in our vision. Similarly, we're at risk of choking with every bite we eat because only a flap separates the passageways for food and air.) Nature's designs represent a tinkerer's improvisations, not an engineer's perfectionism. Simple designs, it turns out, may *not* be nature's choice. That is a hard truth to face, as Francis Crick, perhaps the greatest biologist of the twentieth century, liked to point out. (Crick had begun his scientific career as a physicist.) "Many a young biologist," Crick observed, "has slit his own throat with Ockham's razor."*

That was a modern insight. Isaac Newton had proclaimed that "it is ye perfection of God's works that they are all done with ye greatest simplicity," and in the 1700s, Newton reigned as almost a scientific god.† It counted as a black mark against the microscope that it besmirched the view of science linked with his name. In time, the microscope would once again take on crucial importance. But that moment was far in the future.

---

*Mathematics (and music and chess) are famous for child prodigies. Biologists peak later. "Biology is special that way," says the neuroscientist David Eagleman. "It takes years for people to get a feeling for the organism—for how nature actually works. Young people come in all the time knowing a lot of fancy math. They say, 'What if it's like this computational model, this physical problem?' They're terrific ideas, but they're wrong. Nothing works the way you think it should."

†Newton's contemporaries held him in awe. His masterpiece, *Principia Mathematica*, is still regarded as the greatest of all scientific works. In a poem that accompanied the *Principia*, Edmond Halley wrote, "Nearer the gods no mortal may approach." On Newton's death, in 1727, Alexander Pope composed his famous couplet: "Nature and nature's laws lay hid in night, / God said, 'Let Newton be!' and all was light."

ALONG WITH THESE PRACTICAL AND PSYCHOLOGICAL OBJECTIONS to the microscope came yet another. This time the problem was philosophical. Distressingly, staring through the microscope presented you with discovery upon discovery without ever seeming to bring you nearer to the truth. The hope had been that the microscope would let you see beneath the surface, as if you could see past a clock's face to the cogs and gears behind it.

To everyone's frustration, it turned out not to work that way. You could *see* the cogs and gears, it was true, but you could not see what role they played. To do that, it seemed plain, you had to look closer. Which, with great effort, scientists managed to do. But beneath each level of complexity they found another level just as complex. And each time they managed to bring the new image into focus, the old one slipped from sight, like a suspect disappearing into the fog.

The fear arose that perhaps the entire strategy of looking in ever more detail was misguided. If the questions you were really after were, *What is life? Where does new life come from? Where do babies come from?* then you might map anatomical structures more and more obsessively without drawing any closer to your goal. For modern-day scientists, the analogy might be trying to tell time by looking at the atoms of a clock, or studying the brain ever more minutely in order to learn, *Where does hope come from? Where do ideas come from?*

The brilliant French scientist Blaise Pascal had raised this objection early on. To look through a microscope was to dive into nothingness, he had argued, to plunge through endless layers and never touch bottom. Even inside the body of a mere flea, Pascal wrote, were "legs with joints, veins in its legs, blood in its veins, humors in the blood, corpuscles in the humors, vapors in the corpuscles," and so on, forever. Dive as deep as you like, and would you eventually find treasure? No, replied Pascal, but only "a new abyss."

In the beginning Pascal's pessimism had been a minority view. Then, after days and years spent staring at drops of water and flecks of blood, other scientists began to sound the same dismaying notes. Next John Locke and other philosophers joined in. *If God had meant us to see the hidden world, he would have given us eyes suitable to the task.* Alexander Pope, who had a gift for putting the reigning doctrines of the day into bite-sized form, reproached scientists for not knowing their place. "Why has not Man a microscopic eye? / For this plain reason, Man is not a Fly." The path of wisdom was to recognize that some of God's secrets lay beyond humankind's investigative powers.

THE END CAME QUICKLY. AS EARLY AS 1692, ONLY A FEW DECADES after he had first stared through the microscope, Hooke declared that the game was over. The "subjects to be enquired into are exhausted," he wrote, "and no more is to be done." The microscope, which had been greeted with such high hopes, was fit for nothing but "Diversion and Pastime."

That had left Leeuwenhoek almost alone, except for a handful of amateur naturalists happily examining flower petals and butterfly wings. Leeuwenhoek never wavered in his devotion to the microscope, but after a decade or two of work he conceded that he would not solve the riddle of sex and development. It was not that the riddle had no answer, in his judgment, but that the answer was written in a script so tiny that it would remain forever beyond the range of even the best microscopes. What Leeuwenhoek called the "great secret" would remain a secret. He carried on undaunted, working away even knowing that he would never grasp victory.

Leeuwenhoek's accomplished and melancholy predecessor Swammerdam had reached the same conclusion via a different path. Nature would always hold secrets in reserve because finite humans could not unravel all the mysteries spun by an infinite God. That was disheartening, even though it did testify to God's grandeur. Leeuwenhoek, a more practical and less pious man than Swammerdam, framed things in a cheerier way. He focused his gaze on his rivals around him, not on the heavens above him. He may not have reached his goal, but he had outdistanced all his peers, and that was satisfaction enough.

By the time the grand old man closed his weary eyes for the last time, in 1723, nearly every scientist had set his microscope aside. Confused and adrift, biologists pondered their predicament. The microscope was out, which deprived them of what had seemed their most promising tool. Preformation was out, which deprived them of their guiding theory. Most important, the mechanistic approach was out, which deprived them of a framework with which to make sense of the world.

Where do babies come from? Two thousand years after Aristotle, almost a century after William Harvey, the answer seemed further off than ever.

W ITH THE INTELLECTUAL STRUCTURES THEY HAD TRUSTED NOW collapsed on the ground, even geniuses wandered bewildered through the rubble. The breakdown of the mechanistic view, it soon emerged, was the most disorienting loss. Living creatures are born, and then they grow, and eventually they die. The machine view seemed to throw light only on the last of those phases. Perhaps humans and animals wore out and died much as machines rusted and broke down.

But death and decay were the easy half of the equation; what about life and growth? These were the truly enticing challenges, and the most elusive. "The history of a man for the nine months preceding his birth," the poet Samuel Taylor Coleridge remarked, "would probably be far more interesting than all the three-score-and-ten years that follow it." Even the most familiar examples of life and change presented dark mysteries. When trees grew taller, where did the new wood come from?* What about animals, which not only grew larger but also grew and regrew hair and antlers and other odds and ends? No machines

---

*Photosynthesis was not discovered until around 1779. In one crucial experiment, a Dutch doctor named Jan Ingenhousz put a plant and an unlit candle together inside a transparent, closed container. Several days later, he put a match to the candle, which lit at once and burned nicely. Then he repeated the experiment, but this time with a piece of dark cloth covering the container. After several days in blackness, the candle would *not* light. Ingenhousz concluded—correctly— that when plants absorb sunlight, they breathe out a mysterious substance that sustains candles (and, it would turn out, life).

did anything like that. And those questions arose when you pondered a single, isolated organism. Where were the machines that could make a new machine?

Scientists looked more closely at Fontenelle's mocking comparison of dogs and clocks, from decades before. It was even worse than Fontenelle had said. If animals really were cog-and-gear contrivances like clocks, think how strange those clocks would have to be in order to reproduce. Imagine two parent clocks. To replicate themselves in the way that living creatures do, each parent clock would somehow have to spit out a mini-machine made of scavenged gears and wheels (while continuing to tell the time itself).

Those two whirring, clicking gizmos would join together, every cog and wheel fitting just so, to form a new mini-clock, which would then grow ever larger. (The new clock would grow in two senses: individual gears and wheels would grow larger, and whole new sets of perfectly spinning, turning, meshing gears would somehow come into being.) And, through this entire astonishing process, the new-made clock would have to tick along in perfect order.

No one in the 1700s had ever seen such a machine. Nor has anyone today.

WE HAVE THE GREAT ADVANTAGE OVER OUR INTELLECTUAL forebears, as discussed in the previous chapter, of knowing all about computers and other programmable machines. Without those examples to draw on, growth and development were nigh-to-impenetrable mysteries. But it is worth noting that even from today's vantage point and *with* the wonders of modern technology firmly in mind, the tricks that the body pulls off still retain the power to amaze.

In modern accounts of science, the sections on how organisms grow constantly invoke "recipes" and "blueprints" and "instructions." They should—those analogies are essential. But the familiar words may fool us into forgetting just how bizarre the process of human development (or animal development) truly is. The problem is that we hear the word "recipe" and automatically picture someone in the kitchen measuring and stirring.

But that is to skip past an enormous riddle: Who's following the recipe? It is no great feat to bake a tray of chocolate chip cookies. But imagine if chips and flour and butter had somehow discovered a way *to make themselves* into cookies. "Put simply," writes the biologist Jamie Davies, "cardigans, symphonies, cars, and cathedrals do not build themselves." But living creatures do.

It is even more surprising than that. Every home handyman knows that you turn the electricity off before you try installing a ceiling fan. When you finish, you start the power back up. But living organisms are *always* growing; they don't have the option of shutting down temporarily. And where do the workers come from? It is not a matter of a lone cook in the kitchen. Think of a body's intricate network of nerves and blood vessels. Who are the plumbers who connect mouth and gut? Who are the electricians who wire eyes and brain?

In the kind of building projects we're familiar with, teams of workmen swarm over a house or an airplane while supervisors order them around. But inside the bodies of living animals, no workmen or bosses swoop in from somewhere else. Everything is managed by the body itself, using its own cells both as raw material and as craftsmen. So the cliché that compares difficult tasks to repairing an airplane in midair vastly understates the trick that nature has pulled off. To manage something comparable to what every dog or cat or baby does every day, an airplane would have to rebuild itself while flying, *and* it would have to grow the electricians and engineers who do the work.

All that is only part of the story. Scale makes the tricks that every cell in the body pulls off almost inconceivable. The cells that make up a living creature are invisible to the naked eye—in the human body there are trillions of cells, vastly more than there are stars in the Milky Way—and yet every one of those cells is a chemical factory jammed with pumps and motors and assembly lines, and far more complex than any actual factory.

On top of that, living creatures aren't permanent structures like houses or factories, made of building blocks that rest in place. We are more akin to fountains, say, where the components change perpetually but the pattern endures. It is true that you can never step in the same river twice. But it is not just the river that is always changing.

In a living body, new cells constantly replace old ones. No matter your chronological age, you're formed from parts that are at most a decade old. Scratch your forehead; ten years ago, neither that forehead nor that finger existed. When you meet a friend and say, "You haven't changed a bit," the chemist Addy Pross notes, your friend is in fact almost a completely different person from the one you last encountered.*

Amid such change, what is it that stays constant and makes you *you*? Can it really be that the six-year-old you who ran giggling through a garden sprinkler is the same person as the stiff and stodgy coot fumbling for his reading glasses? It seems both indisputable and unfathomable. In learned journals today, philosophers wrestle endlessly with such riddles. After a kidney transplant you would still be yourself. What about after a brain transplant? Would you be yourself, or would you be *him* wearing your body like a borrowed overcoat?

In the mid-1700s the details of such questions about personal identity had not yet emerged, but the makers of the modern age had glimpsed the complexity of the living world. Astonished and bewildered, they cast about for some model of how development could possibly take place.

They began with a giant step backward.

---

*Plutarch explored this puzzle two thousand years ago, in his parable about Theseus's ship. Theseus was the hero who slew the minotaur. For a thousand years afterward, grateful Athenians preserved his ship, replacing each plank with an identical copy when worms and weather had taken their toll. Eventually not an original board remained. Was the rebuilt ship a new ship altogether, Plutarch asked, or was it still Theseus's ship? (If it *was* new, at what point had the old ship vanished? And if it was *not* new, did it make a difference that Theseus had never set foot aboard "his" ship?)

EIGHTEEN

A VASE IN SILHOUETTE

P ERHAPS THE CLOCKWORK ANALOGY DID NOT APPLY TO LIVING creatures, the new generation of thinkers conceded, but that did not mean they had to give up on scientific explanations altogether. They began by wrapping themselves in the cloak of the revered Isaac Newton.

Pierre Louis Maupertuis, the French astronomer-turned-biologist who delighted in mocking the preformationists, was the first to make the new case. Newton and his successors, Maupertuis proclaimed in 1745, had shown how gravity explained the structure of the solar system. Now the time had come to "extend it further than the astronomers have done. Why if this force exists in Nature, should it not have a part in the formation of animal bodies?"

Maupertuis's invocation of Newton and gravity was new. Oddly, he paired that new idea with an old one. Nearly a century before, scientists had rejected the theory that there were two semens, one male and one female. Maupertuis not only signed onto that long discredited view but promoted it enthusiastically. Male and female semen contain tiny particles that come from every part of each parent's body, he explained. When a man and woman have sex, those "seminal particles" meet in the uterus, where gravity draws together particles from corresponding organs in the two parents, thus forming new organs that carry traits from both partners.

This was a bold—or far-fetched—suggestion, with several virtues. It explained why children could resemble either parent (or both), which preformation had conspicuously failed to do. Just as important, it moved away from the clockwork model but retained its most important element. Though cogs and gears were banished, Maupertuis's picture still evoked the orderly, law-governed workings of the solar system, the grandest machine of them all.

As a bonus, Maupertuis managed to turn one of the oddest features of Newtonian science to his advantage. Newton's picture of a perfectly regulated cosmos depended entirely on the all-embracing power of gravity, but Newton had never explained what gravity is. "Ye cause of gravity is what I do not pretend to know," he acknowledged, and so he confined himself to describing the effects of that utterly mysterious force. Somehow it reached to the farthest corners of space, instantaneously, exerting its pull across billions of empty miles. From the most distant star to the nearest flower, every object in the universe felt gravity's tug. Why shouldn't gravity also draw the seminal particles together deep within a woman's body?

In the judgment of posterity (though not of his own contemporaries), Maupertuis scored other coups, too. His gravitational theory explained why males and females of different species rarely breed (because the parents' seminal particles are not enough alike) and why birth defects are not outlandish mistakes, like a foot growing where an ear belongs, but predictable errors, like six fingers rather than five (because gravity draws similar particles together). In some ways, Maupertuis even managed to anticipate Gregor Mendel's account of genetics. Maupertuis suggested that particles can remain dormant for several generations, for instance, which would explain why a child of brown-haired parents might have flaming-red hair like her grandmother. And he suggested that particles might suddenly change, giving rise to new traits that later generations of scientists would dub "mutations."

Claiming Newton's support was good strategy, and Maupertuis quickly won allies. The most important was Georges-Louis Leclerc, later known as the Count de Buffon, a French scientist and a peacock in human form. (We briefly encountered Buffon once before, mocking the preformationists with a calculation that showed that Russian dolls would quickly shrink to microscopic size.) One of the most admired thinkers of the age, Buffon had nothing in common with the unworldly, socially inept caricatures who already featured in every popular account of science. "He loved money and became rich," one biographer tells us. "He loved power, and he frequented those in power. . . . He loved women, and not just for their beautiful souls. His laboratory experiments were few and his assumptions often questionable. He let his imagination go well beyond the facts." He was also, indisputably, both learned and brilliant.

Buffon began his career as a mathematician and a kind of European ambassador for Newtonian science. But he soared to fame on the strength of a monumental, thirty-six-volume work called *Natural History* that appeared over the course of his long life. These unlikely best sellers ranged over a multitude of subjects—mammals, birds, human beings, geology, anthropology—in lively and opinionated fashion. "Sloths are the lowest term of existence in the order of animals with flesh and blood," Buffon noted. "One more defect would have made their existence impossible."

The public grabbed up Buffon's books. He soared past Voltaire and Rousseau in renown, and his every pronouncement carried weight. (Thomas Jefferson grew so furious at Buffon's claim that all forms of life in the New World were small and feeble that he instructed Lewis and Clark to keep their eyes out for mammoths and mastodons, which he believed still roamed the wild.) Buffon happily played the part of the great man: he favored silk waistcoats and lace cuffs and had his hair curled several times a day; he employed a renowned chef and presided at long dinners, where the sparkling conversation veered often to the scandalous; he spoke fondly and frequently of his many accomplishments. When he died, in 1788, fourteen liveried horses and a thirty-six-member choir led the funeral procession, and twenty thousand spectators along the parade route scrambled for a view.

So when this titan of science weighed in on sex and conception, everyone listened. Buffon began, like Maupertuis, by rejecting the notion that women have eggs. That long-held view, he declared magisterially, was "badly founded" and "explained nothing." He noted, correctly, that no one had actually seen mammalian eggs but only the ruptured follicles that had presumably once contained the eggs. For de Graaf and his fellow anatomists, the presence of those burst follicles in the ovaries had testified unmistakably to the presence of eggs; this was hardly more of a leap than seeing cracked and empty eggshells on a kitchen counter and deducing that someone had made breakfast.

Buffon didn't buy it. All that de Graaf had been entitled to conclude was that *something* had emerged from the follicles. De Graaf believed that mysterious something was an egg. Wrong, said Buffon. It was female semen.

This was considered racy, which only drew more notice. Buffon thrived on the controversy. His prose style was ornate, but he enjoyed hinting that his adversaries were scolds and prudes. (Buffon's foes took his elegant prose as proof that his ideas were lightweight and unworthy. That he was French besides seemed to them further evidence of their charges.) Women could only produce the *liqueur seminale* that made conception possible, Buffon made a point of emphasizing, if they enjoyed themselves during sex.

In Buffon's scenario, semen was "in both sexes, a sort of extract from all body parts." When male and female extracts mixed, they produced "a sort of rough sketch of the animal, a small organized body in which only the essential parts are formed." That was vague, and Buffon's account of how it happened was vaguer still. Somehow gravity or some other force drew together bits that "need one another."

This theory was closely related to the one that Maupertuis had proposed a few years earlier. Both men spoke of two semens; both talked of Isaac Newton and mysterious forces of attraction. But Maupertuis had presented his argument in an eccentric, quasi-erotic book called *The Earthly Venus*, which he published anonymously. (One modern historian describes it as a "serious physical argument" scattered within a "symphony of witty and learned smut.") Buffon argued his case in an imposing volume that all Europe clamored to read. Maupertuis whispered, and Buffon roared.

The details in these new theories were not important. That was fortunate, since Buffon and Maupertuis had scarcely provided any details at all. What was important was that they had managed to jolt the sex and babies debate out of its old rut. Others before them had pointed out flaws in the preformationist theory—mules, monsters, family resemblances—but Buffon and Maupertuis had moved beyond criticizing. They had proposed an alternative, albeit a rudimentary one.

The secret of development, in their view, was not the unveiling of Russian dolls made at the dawn of time on God's assembly line. Instead, living organisms not only grew but changed, by adding countless new structures that appeared from scratch. This was an old idea, as we have seen, with such formidable advocates as Aristotle and William Harvey. But in the past, developmental theories had been dismissed as unscientific and implausible—*Do you mean to say that babies grow by magic? And is it your view that complicated structures like hands and eyes arise from nothing at all?* Now, in the mid-1700s, Aristotle was out of fashion, and Harvey was long dead. Maupertuis was controversial but widely admired, and Buffon was the man of the hour.

With its allusions to mysterious particles tugged by gravity-like forces and somehow patted into shape, the new picture of development was more a suggestion than a proper theory. These were stumbling steps. But they were steps that headed off in a long-neglected direction (and in the direction that would prove, centuries later, to be the correct one). Buffon and Maupertuis had not solved the baby riddle, but they had revived an old idea and given it new respectability. Perhaps in time someone would build on their ideas and take them further.

INSTEAD, A LONG-SIMMERING BATTLE BROKE OUT INTO OPEN fighting. The differences between the scientists who espoused the Russian doll theory and those who favored the new, developmental model—between preformation and epigenesis, in the language the scientists used—seemed stark and momentous. The two sides flung angry charges back and forth.

Why did the preformationists insist that bodies do not change as they grow, except in size? That was plainly false, cried their rivals. The body can grow new bits, not just unroll preexisting ones, and they cited countless everyday examples. There was not even any need to venture into the charged territory of embryos and babies. Think of warts or tumors or moles, which appear out of nowhere. Think of how cuts heal, as new tissue mends tears in the flesh. Think of scars, which surely were not predestined.

The preformationists fought back with vigor and disdain. For years they had railed against the notion of a "vital force" that supposedly guided living creatures on their way. This talk of mysterious guiding forces was vague and empty; it conjured up images of a puppet master working invisible strings. All such theories, one preformationist charged, were "trash, regurgitations of occultism, irrationalism, and pseudo-science."

And if the advocates of epigenesis maintained that somehow organisms grew *without* guidance, then how could they deny that their system relied on chance, that dangerous and damnable notion? If intricate new structures arose out of the blue, how did living creatures take on such perfect, complex form? Worse yet, the preformationists thundered, this dangerous doctrine threatened to push God out of the picture. "Beware," one scientist warned, "that it is very dangerous to admit the formation of a finger by chance. If a finger can form itself, a hand will form itself, and an arm, and a man."

Each charge drew a countercharge. The accusation that they believed in chance drew special fire from the epigenesis camp. *False!*, they shouted. Their opponents talked as if change and chance were the same thing. But everyone acknowledged that the body changes as if it were following a script. The merest glance tells us which act of the drama we are witnessing. What could be more predictable than the path from helpless baby to coltish teen to adult in full command of her powers?

Some changes are less welcome but roll out inexorably even so—backs stiffen, hair grays, skin wrinkles. These are not the nicks and dings of hard wear, akin to the scars that a kitchen table accumulates over the years, but built-in, universal features of life. None of this fit with the picture of life arising from miniature dolls.

F ROM OUR POINT OF VIEW, DOWNSTREAM IN TIME, THE TWO SIDES appear closer than anyone could see at the time. Both had latched onto important truths. There *is* something preformed in an embryo, in the sense that development unfolds according to instructions that were written down at conception. And there *is* something more going on than unfolding, because completely new parts—not just miniatures grown big—do arise at specified junctures in the timetable.

Maddeningly, this was not a dispute that could be resolved by careful observation. No matter how meticulously you tracked new-laid chicken eggs, say, you could not be sure if heart or brain or beak had appeared one day where they had never been before, or if they had been present all along but in so tiny (or transparent or distorted) a form that they could not be recognized. Instead, each side told its own story over again and crafted new insults for its rivals.

Between catcalls, both sides grudgingly shored up their own arguments. By the time they had finished adding caveats and concessions, they had gone a long way to bridging the gulf between them. The preformationists' Russian dolls turned out to be unlike any dolls anyone had ever seen, so prone were they to changing shape and proportion. And the epigenics camp had backtracked and improvised, too. New structures did not truly materialize, they suggested, but grew out of buds or germs or precursors.

At the time, no one saw that the opposing factions had stumbled toward common ground. "What is the difference between saying that the whole organism preexists, but not in the form in which it will later appear," asks one modern historian, in exasperation, "and saying that it exists only potentially, or that some precursors of its parts are present?"

Today's textbooks tend to cast the tale as a battle between the far-seeing epigenesis side, who had the story nearly right, and their misguided rivals in the preformationist camp, who clung to a complicated and far-fetched doctrine. But though the preformationists lost out, their views were not as crazy as all that. We now know that on the day she is born—years before any thought of pregnancy will cross her mind—an infant girl carries within her body more than a lifetime's quota of eggs, around one million altogether. (Men do not produce sperm until puberty.) That is a long way from saying that she carries an endless supply of miniature humans nested inside one another. But perhaps it is enough of a step in that direction to make us less inclined to deride our forebears for their foolishness.

And the notion of an endless sequence of preformed miniatures veered near a crucial insight. Within every sperm and egg, we now know, is a double helix of DNA that serves as blueprint (actually, half the blueprint) for any child the parents conceive. In time that child may grow up and pass along her own DNA to a child, who may someday pass along *her* DNA, and so on. A long line of messages in cells is not the same as a sequence of dolls within dolls, though clearly there is some resemblance. The preformationists had thought too literally, but they had thought deeply.

These near misses might make us think of the blind men and the elephant, but that is the wrong analogy. The blind men's problem was that they had each grabbed a different part of the elephant and taken it for the whole beast. Here those who believed in preformation and those who favored epigenesis had hold of the same thing—the observation that living creatures somehow "know" how to grow—but both sides insisted that only their opinion was valid. Rather than an elephant, we might think of one of those drawings that show simultaneously a vase and faces in profile. "It's a vase," shouted the preformationists, and "No, it's two faces," their rivals shouted back.

The shouting and skirmishing had gone on for almost a century, but a breakthrough was near at hand.

N O ONE COULD HAVE FORESEEN THAT PROGRESS WOULD COME not from a new insight or a shocking discovery, but from the revival of an ancient and deeply mistaken belief. This irrepressible notion was dubbed "spontaneous generation": the idea was that life could rise up, on its own, from a heap of rags or a lump of rotten meat. Deep inside a neglected bit of filth, flies or fleas or toads might stir themselves to creeping, hopping life.

The idea had no merit. But it had appeared in so many guises through the millennia, after having been knocked down and left for dead time after time, that it seemed to be itself an example of how life could arise from nonlife. Now Buffon and Maupertuis had revived the old belief once again. This time, though, an array of great thinkers and experimenters felt moved to shoot down the spontaneous generation claims once and for all. And it was those efforts to prove where life did *not* come from that would, in the end, finally point the way to where it did come from.

Foremost among these talented skeptics was an Italian priest and scientist with a reputation for designing original, but deeply odd, experiments. Now he would outdo himself.

PART FOUR

THE CLOCKWORK TOPPLES
AND A NEW THEORY RISES

*It takes a thousand men to invent a telegraph, or a steam engine,
or a phonograph, or a telephone or any other important thing—
and the last man gets the credit and we forget the others.*

— MARK TWAIN

NINETEEN

FROGS IN SILK PANTS

E VEN IN AN AGE OF ALL-AROUNDERS RATHER THAN SPECIALISTS, Lazzaro Spallanzani's versatility set him apart. He was a mountain climber, a bold traveler who ranged as far afield as a sultan's harem in Istanbul, a mathematician, a Greek scholar (his first publication was on *The Iliad*), and, most important, by one historian's reckoning, "undoubtedly one of the greatest experimental biologists of all time."

It had seemed, early on, that his career would be more routine. In 1749 he had begun studying law at the University of Bologna, in line with his father's wishes. Here he came under the sway of a remarkable woman who happened to be a cousin on his father's side. Laura Bassi was a mathematician and a physicist, already famous as the first woman in Europe to hold a university professorship. Bassi recruited her young cousin to science (and convinced Spallanzani's father to accept the decision).

Once launched, Spallanzani never slowed down. He was a man of serial obsessions, and he poked his beaky nose and his dark eyes into everything. The living world beckoned him—*How do bats navigate? How do deep-sea fish glow?*—and so did the inanimate world—*Why do thunderclouds form? What causes earthquakes?* On the trail of a mystery, he made Inspector Javert seem like a slacker. Spallanzani once clambered up an erupting volcano, so close to the flowing lava that the fumes knocked him unconscious. (He wanted to know how fast lava travels.) Determined to learn how bats find their way, he methodically worked through each sense in turn, outfitting his poor captive bats with blinders, then with earmuffs, with nose plugs, and so on, to see if they could fly through a room bristling with obstacles.

To see if snails could regrow their heads, he decapitated *seven hundred* of them. To see why stones skip across water, he flung thousands upon thousands and wrote a detailed mathematical analysis of how they bounce. To study digestion, he hid scraps of meat in linen bags, lowered them down the throats of turkeys and owls and frogs and newts and snakes and cats and dogs (and, with "some apprehension," one human, Spallanzani himself), and then hauled them up again to have a look. To see if the body's heat made food break down faster, he crammed bits of meat and crumbs of bread into glass tubes and kept the capsules tucked under his arms for two days. So formidable were his skills as an experimenter that Spallanzani's peers dubbed him the "Magnifico."

His experiments with bats illustrate his persistence, his openness to even the unlikeliest possibilities, and his ingenuity. (They reveal, too, a casual willingness to inflict pain on animals that was a standard feature of his era.)* Spallanzani's curiosity had been roused in the first place by watching a tame owl. The bird could find its way perfectly by the light of a single candle, but not in the dark. Bats could fly in the dead of night, though, and in the blackest caves. Could it be that bats had better night-vision than owls?

Spallanzani caught three bats in a cave and brought them indoors, into a small room that he had made into a kind of obstacle course. Strings dangled from the ceiling, wide enough apart for a bat with outstretched wings to fly through. Strips of metal foil or tiny bells hung on each string and sounded if a bat brushed by. On a moonless

---

*Spallanzani noted indignantly, in the course of his digestion experiments, that eagles, falcons, and dogs "used all their efforts to bite me." Good for them!

night, Spallanzani, his brother Niccolo, and a cousin sat silently in the gloom. Twenty minutes of candlelight, then twenty minutes in the pitch-black, the cycle repeated over and over, the only sounds the occasional tinkling of a bell as a bat passed near a string and the beating of wings coming nearer, then receding, then returning.

To test whether vision was crucial, Spallanzani blocked his bats' eyes with small, opaque discs. Blind as bats, they nonetheless flew with their customary grace. Next, taste. He cut out their tongues. No problem (for Spallanzani, at any rate). Onto touch. Perhaps bats could feel tiny air currents bouncing against their bodies from obstacles as they flew by? Spallanzani painted his bats with shellac. Despite what he called a "light varnishing" (and then a second coat, and even a third), they flew perfectly. One unfortunate bat even navigated its way despite a messy coating of paste and flour. Now, smell. Spallanzani blocked the bats' nostrils. Away they flew.

Finally, sound. Spallanzani made tiny cones out of brass, shaped like cheerleaders' megaphones, and wriggled them into the bats' ears. Sporting their strange headgear, the bats flew along contentedly. (This first test was just a precaution. It might have been that the cones were so cumbersome that the bats couldn't tolerate them.) Then came the key move: Spallanzani blocked the cones with tar, so that the bats couldn't hear. This time the bats collided with one another, smacked into obstacles, and even crashed into the ground.

These were astonishing experiments, and they won Spallanzani scorn as well as praise. "Since bats see with their ears," another scientist mocked, "do they hear with their eyes?" Spallanzani had not sorted out the full story.* The great mystery he left unraveled was that bats could navigate by hearing even though they made no sound.

---

*Spallanzani was 150 years ahead of his time, but he had missed a crucial insight. He believed that bats have super-acute hearing, which let them detect the buzz from flying insects or the sound of air from their own wings bouncing off nearby surfaces. The true story, which emerged at around the time of World War II, was that bats emit high-frequency peeps and chirps and listen for the echoes. When biologists first gave a talk explaining the bats' technique, research on radar and sonar was still top secret; the bat researchers were accosted by scientists who were outraged that they had divulged military secrets. But bats had figured it out 50 million years before human engineers.

What could they be listening to? Even so, Spallanzani had made giant headway on an ancient riddle. He had done it by devising experiments that few others would have thought of—*bats in batter? bats wearing ear cones?*—and by taking his findings seriously, no matter how unlikely they might be.

Those were the very traits that would bring him nearer to the truth than anyone had come before, when he took on biology's oldest, deepest riddle. It would be Lazzaro Spallanzani who took the next giant step toward solving the mystery of fertilization. On history's calendar, all this took place only yesterday. Spallanzani was a contemporary of George Washington and Thomas Jefferson. (In 1776, when Jefferson sat in Philadelphia writing the Declaration of Independence, Spallanzani was at work in Pavia, Italy, writing *New Observations and Experiments Concerning the Spermatic Animalcules*.)

In the era of the founding fathers, no one in the world knew how fatherhood worked.

SPALLANZANI CAME TO THE STUDY OF SPERM AND EGGS BY WAY OF another of the age's great preoccupations. This was the doctrine of spontaneous generation, which had been left for dead in the late 1600s but revived by Buffon and others in the mid-1700s. Perhaps that resurrection could have been foreseen, for a long, long roster of thinkers through the ages had argued vehemently that, in the right circumstances, living creatures might emerge from lifeless odds and ends. Deep into the 1600s, even the most eminent scientists took for granted that life—especially life in its nastiest forms—might climb out of nearly any dark corner. The most admired of all scientists, Isaac Newton, believed in spontaneous generation. So did the famously skeptical philosopher René Descartes. "So little is necessary to make an animal," Descartes remarked, that it was no wonder that rodents, worms, and bugs arose spontaneously from dead flesh.

The more lowly or despised the animal, the more likely that it popped into existence on its own. It was well known, for instance, that spiders emerged from rotting mushrooms. Both William Harvey and Robert Hooke reported, after careful investigation, that insects arose from dying plants. At a meeting of the Royal Society on June 24, 1663, Hooke was given the task of examining "viper-powder"—dried, ground-up snake—under a microscope to test eyewitness reports that "a box of viper-powder, which being opened and found extremely stinking, had [a myriad of] little moving creatures in it, like mites of cheese."

A year later, the Society still had not settled the question. Sir Robert Moray passed along an eyewitness account of "a pot with viper-powder in it, brought from Venice," and kept tightly sealed. Six months later, it was "full of little live insects." Another member of the Royal Society chimed in: "He had known a chemist who used to perfume his viper-powder with myrrh, to preserve it from breeding worms."

What scientists believed they had learned from the close observation of nature, ordinary people took for granted. No prudent person, warned a seventeenth-century English physician, should consume creatures that originated in "the excrements of the earth, the slime and scum of the water, the superfluity of the woods, and the putrefaction of the sea: to wit . . . frogs, snails, mushrooms, and oysters." The English, who had long derided the French and Italians for their dining habits, now claimed that they had sound medical reasons for their disdain.

Shakespeare reflected the same folk beliefs. Crocodiles creep forth from "Nilus's slime," he remarked in *Antony and Cleopatra*, and in *Hamlet* he commented matter-of-factly that "the sun breeds maggots in a dead dog." This was little more than common sense. A vast gulf separated majestic creatures like lions and tigers and, of course, humans, from vermin and other lowly beasts. It would be no surprise if those debased forms of life originated in some way that matched their position in life's cellar.

Religion, it had long been assumed, taught the same moral. Noah did not bother to bring pairs of mice and flies and similar creatures aboard the ark, theologians had explained ever since Saint Augustine, because there was no need. Those ignoble animals would turn up on their own, spontaneously.

Finally, in 1667, the debate had shifted. Francesco Redi, personal physician to the Medicis in Florence, had carried out a series of experiments on the origin of life that are still hailed today for their simple, clear design. Redi was a dazzler. He was elegantly slim (one admirer described him as "the picture of hunger"), effortlessly articulate, and a brilliant scientist as well as a witty and charming courtier. He wrote poetry (his ode to Tuscany's wines is still read); he described his scientific ventures in chatty, engaging prose; he flattered with so light a touch that the praise seemed sincere rather than unctuous. Enthralled by Redi's research and entranced by his manner, Archduke Ferdinand II indulged his scientific pet in all his far-ranging ventures.

Redi had risen to fame by sorting out, in response to a request from the archduke, how venomous snakes do their damage. Snakes were a common hazard in Tuscany, but no one knew how they manufactured their venom or how it worked. Redi demonstrated that snakes killed their victims by injection. Venom did not work like poison; a person who swallowed venom would walk away unharmed. (Redi, who understood the Medici fondness for showmanship, enlisted the services of the royal snake catcher to show what he had learned. That intrepid fellow poured a spoonful of venom into a glass of wine and drank it, Redi noted, "as though it were some pearly julep." Then he licked the spoon.)

When Redi turned his attention to spontaneous generation, his findings were more important, but the show was not as engaging. Instead of an imperturbable snake catcher wrapped in writhing serpents, Redi's props were festering hunks of meat. Redi set out boxes of meat scraps in the sun, some of the boxes covered with a gauze mesh and others identically prepared but open to the air. Within a few days, he found flies inside the open boxes and worms "creeping up, all soft and slimy," but no maggots or flies in the covered boxes.

Life came from life! Flies did not arise from dead flesh but from eggs laid by other flies. Redi's experiments rank as landmarks in the history of science. Still, his findings did not mark the end of the controversy over spontaneous generation, because its proponents could always come up with new candidate creatures. *Okay, not flies, but what about . . . ?*

The trend through the years was that the proposed organisms grew steadily smaller and less imposing. In the mid-1600s, well-regarded scientists had published recipes for producing mice. (The key ingredients were a sweat-soaked shirt and a few grains of wheat. Presumably it was the wheat that confused matters, by luring mice that had been created in the usual fashion.) Later in the century, the focus shifted to insects and flies. Later still, when the microscope came along, the discovery of countless new forms of life revived the old debate.

I N THE MID-1700S, SPONTANEOUS GENERATION ROSE TO PROMInence yet again. Two men in particular, our old friend the Count de Buffon and an English priest-turned-scientist named John Turberville Needham, fueled the rise. The two scientists collaborated, Buffon sticking mostly to writing and Needham to the microscope, both of them convinced that nature possessed a "vegetative force" that could bring matter to life. In a series of famous experiments in 1748, Needham boiled mutton broth and then poured the steaming liquid into glass flasks and sealed them. If any microorganisms appeared, Needham announced, that would prove that life could arise spontaneously. Days later, he found the broth teeming with microorganisms. Life had appeared, out of nowhere, in a sterile soup!

Spallanzani took up that challenge. In a series of experiments akin to Redi's, Spallanzani boiled broth and poured some into flasks that sat open to the air; he poured identical batches of broth into identical flasks, and sealed the second set of flasks. Then he sat back to watch for signs of life.

The key was how Spallanzani's approach differed from Needham's. Instead of boiling his samples for ten minutes, he boiled them for an hour. Instead of sealing his flasks with corks, he melted the necks over a flame, which made them airtight.* In every case, broth exposed to the air (or in flasks sealed only with a cork) quickly swam with microorganisms. Broth in sealed flasks remained lifeless.

Spallanzani, an even-tempered man who permitted himself an occasional sharp-edged remark, proclaimed victory. The newfangled "vegetative force" was nothing but the old, foolish doctrine of spontaneous generation in disguise. This was old wine, gone bad, in a new bottle. Satisfied that he had dealt a death blow to spontaneous generation, Spallanzani now turned to a closely related but even more ambitious question. If life arises only from life, exactly how does that work? In particular, Spallanzani wanted to know, what does semen have to do with fertilization?

THIS WAS STILL A QUESTION MIRED IN DARKNESS. THE LEADING view, that semen worked its influence by "aura" or "emanation," dated back more than a century to William Harvey (and, in a different form, all the way back to Aristotle). The rival view focused on the "animalcules" Leeuwenhoek had found swimming by the millions in semen, some eighty years before. Those tiny swimmers were the secret to life. Somehow they contained a miniature of the not-yet-developed organism. Ejaculated by the male, they swam into the female, where one of the millions of racers burrowed into the uterus and started to grow. That was Leeuwenhoek's view (he made no allowance for the egg, as we have seen). But, put off by the incredible number of sperm cells and their wormy appearance, nearly everyone else disagreed.

---

*A century later, Louis Pasteur would add a refinement by leaving the flask's neck open to the air but shaped into an ever-so-thin curve like a swan's neck. The curved neck played two roles; its shape meant that nothing could fall from the air into the broth, and the opening at its end spoke to the objection that perhaps microorganisms would have arisen spontaneously if they'd had access to fresh air.

Spallanzani's first inspiration was to use frogs as his experimental subjects, because they shed their eggs outside the body, which brought the fertilization process out into the open, where it was easy to see. This was not as straightforward as it sounds. The great Swedish naturalist Carl Linnaeus had proclaimed, with characteristic certainty, that "in Nature, in no case, in any living body, does fecundation or impregnation of the egg take place outside the body of the mother."

To contradict so eminent a figure took daring. Linnaeus was consumed with two subjects above all. The first was finding order within the endless variety of the natural world. The second was his own magnificence. "God creates, Linnaeus arranges," he boasted. He once commissioned an engraving of Apollo for a botanical work and instructed the artist to depict the Greek god with Linnaeus's face. But Spallanzani, an odd mix of swashbuckling adventurer and detail-driven perfectionist, was temperamentally not much inclined to defer to even the most august authority.

Countless hours of observation soon made him expert on what he called "the *amours*" of the frog. (He also studied what one historian called "the nuptials of the newt.") The male frog clasps the female from the back, she exudes a stream of eggs into the water, he releases his semen on them, and the partners swim apart. (Mating males clasp their partners so tenaciously that Spallanzani found it almost impossible to pry them away. Not even amputation—first of one limb, then of two, and then of the male's head!—dampened their enthusiasm.)

This certainly looked like external fertilization, contrary to Linnaeus, but Spallanzani was a man who dotted every *i* and then went back and polished the dots. It was hard to be sure just what was happening in all that water, especially with the female "darting backward and forward" and the male "throwing himself into strange contortions," and both of them croaking away all the while. Spallanzani put pairs of frogs into empty, dry containers, where it was easier to keep watch. Even in this unfamiliar setting, the partners carried on undaunted, the males releasing "a small jet of limpid liquor" upon the eggs. Spallanzani put

FIGURE 19.1. This statue of Spallanzani stands in the town square in his hometown, Scandiano, in northern Italy. The frog is not wearing pants.

those eggs into water and kept watch. They grew into tadpoles, which grew into frogs. In frogs, fertilization takes place outside the body. That took care of Linnaeus.

With the preliminary obstacles cleared from his path, Spallanzani marched decisively ahead. But in quest of the biggest prize of the age, the big-game hunter could hardly have looked less imposing. The hero's weapons were not rifles or spears but needle and thread. Spallanzani had assigned himself a task that sounds like the sort of absurd challenge imposed, in a fairy tale, by a cruel king on a hapless prisoner. He sat at his workbench with cramped fingers and weary eyes, cutting and sewing dozens and dozens of tight-fitting, miniature boxer shorts made of silk. For frogs.

The point of the boxers was to prevent the male's semen from reaching the female's eggs. Would the females become pregnant even so, as the "seminal aura" sent out its ghostly waves? Or would the shorts, which were wax-coated as an additional safeguard, serve as a full-body condom?

Spallanzani did not describe the boxers in any detail, and though he was a skilled artist, he made no drawings (it is tempting to picture the shorts as adorned with hearts or even with frogs). "The idea of the breeches, however whimsical and ridiculous it may appear, did not displease me," he wrote gamely, "and I resolved to put it into practice."* He wrestled the males into their outfits. Undeterred, they sought out the females with their customary eagerness, Spallanzani wrote, "and performed, as well as they could, the act of generation."

Then he gathered up the eggs. Half came from females who had mated with boxer-clad males, half from females whose partners had carried on au naturel. Spallanzani peered at the two sets of eggs. Which would grow into tadpoles?

---

*Spallanzani owed the idea to two French scientists who had tried unsuccessfully to fit frogs with underwear in 1740. Years afterward they told him of their mishaps in a letter. Either the shorts were so loose that the frogs wriggled free or so tight that they could scarcely breathe, let alone mate.

TWENTY

A DROP OF VENOM

S PALLANZANI SOON HAD HIS ANSWER: THE EGGS THAT HAD BEEN doused with semen by naked, unencumbered males developed into frogs; the eggs that semen had not touched did not. Decisive as that seemed, Spallanzani carried out a barrage of follow-ups. He began with the besmirched boxers. He gathered a tiny bit of semen from the cloth and daubed it on some eggs. Those eggs developed into frogs. Untouched eggs did not develop.

This was a milestone in the history of science. Here was an experiment so simple that it could be understood at a glance and that nonetheless, and in one bold coup, resolved an age-old question. *Yes,* Spallanzani's demonstration proclaimed, *semen and egg are both necessary if fertilization is to occur, and they have to make contact. The "seminal aura" was a myth.* In a Museum of Science that featured dioramas of landmark moments, the wing devoted to the 1600s and 1700s might depict Newton conked on the head by an apple (which certainly never happened), Galileo lugging rocks to the top of the Leaning Tower of Pisa (which quite likely never happened), and Spallanzani tightening the waistband of his frogs' boxers (which absolutely did happen).

But there is a twist in the tale. Considering how odd it is, perhaps it counts as a full-fledged knot. The twist is this: ever since Spallanzani's day, historians have lauded him as the man who proved beyond doubt that, in the story of sex and conception, sperm and egg are equal players. But Spallanzani never believed that for a minute. Instead, he interpreted his own experiments in a way that later scientists ignored or never even suspected. Convinced that Spallanzani's marvelous discoveries spoke for themselves, his intellectual descendants credited him with finding truths that he explicitly disavowed. This is a bizarre fate, as if a lifelong pacifist like Gandhi found himself hailed by generations of West Pointers for his brilliance as a military strategist.

What is most striking, to our eyes, is that Spallanzani focused nearly all his attention on semen and almost none of it on the hordes of spermatozoa swimming within that semen. He waved aside those countless swimmers as little more than distractions, probably parasites that happened to live in seminal fluid. Semen, he believed, had some crucial property that sparked eggs to life. Most likely it was chemical. Oddly, he made an analogy to snake venom. By mysterious means, venom had the power to trigger a cascade of dangerous changes in the body; semen was a benign counterpart, a mysterious fluid that triggered life rather than death.

But for so meticulous a scientist as Spallanzani, the first task was to prove beyond the possibility of doubt that semen induced fertilization by direct, physical means rather than by "contagion." In the experiment where he took semen from the frogs' underwear and painted it on eggs, Spallanzani had essentially created test-tube frogs. That was a first, and a first that seemed to render any second unnecessary. But the "priest cum scientist," as one historian described him, was only getting started. He extracted semen from the seminal vesicles of male frogs and painted it on new-laid eggs. Again, tadpoles! (And again, from untouched eggs, nothing.)

Next Spallanzani turned to species where fertilization is internal. In a remarkable display of microsurgery, he extracted semen from male silkworms and spread it on silkworm eggs. Soon, moths! (And no signs of fertilization in untouched eggs.) He tried dogs. He borrowed a spaniel in heat, locked her away from other dogs, and squirted a syringe of semen from a male spaniel inside her. Sixty-two days later, three puppies! (Spallanzani noted without comment that the pups "resembled in color and shape not the bitch only, but the dog also from which the seed had been taken.")

So semen had to make actual contact with the egg, which seemed to imply that it had some special properties. But perhaps that was too bold a leap. Could other substances also induce fertilization? Persistent to an almost unfathomable degree, Spallanzani gathered up droppers and scalpels and set to work. He began with a bit of blood from a frog's heart and dabbed it on some eggs, to see if anything happened. *Nope*. (The rationale was that a beating heart was practically an emblem of life.) Nor did any other juice extracted from a frog's heart have any effect. Neither did extracts from lung or liver or doses of vinegar or wine or urine, in various combinations and dilutions. Nor did lemon juice or lime juice, or extracts of oil from their peels. Nor did a jolt of electricity. Somehow, semen was special.

In keeping with that line of thought and with his picture of semen-as-chemical-potion very much in mind, Spallanzani embarked on a new round of experiments. This time his goal was to see if semen retained its fertilizing power even if it was diluted in water. To his amazement, it did, almost regardless of how far he pushed things. Diluting and diluting and diluting and then launching into seas of calculation by way of trying to find *some* limit to semen's power, Spallanzani found himself staring at ever more incomprehensible results. He took a frog egg and a drop of water from a beaker of water with a tiny bit of semen stirred in. "I found the volume of the egg to the volume of the spermatic particles as 1,064,777,777 to one."

This was astonishing, but Spallanzani's chemical model almost forced him to miss the point. If he had taken spermatozoa more seriously and recognized that it takes only a single sperm cell to fertilize an egg, his finding would have made perfect sense. He chose, instead, to try a variant of his dilution experiments. Once again, he would knock his head against a piece of low-hanging fruit. Once again, he would ignore it except to complain that someone needed to tidy up this orchard.

This second go-round relied on filters. Spallanzani dissolved frog semen in water and passed the mixture through a piece of paper. The filtered material still fertilized eggs, although not quite so well as an unfiltered mixture. The same held for a semen and water mixture poured through two sheets of paper, or three. Finally, by the time Spallanzani had poured the mixture through half a dozen sheets, it no longer fertilized eggs. But when Spallanzani took a dab of the goopy residue left behind on the stack of filter papers, mixed it in water, and painted that semen and water mix on the eggs, he ended up with as many fertilized eggs as he'd ever had.

The conclusion, in Spallanzani's mind? Some component of semen carried its fertilizing power, and it was that component that ended up caught on the filter paper. That fit perfectly with his chemical model, and Spallanzani cheerfully recorded his findings. He never put the viscous blob on his filter paper under a microscope. If he'd looked, he would have seen it crowded with wriggling sperm cells. Perhaps he'd have wondered why.

DISMISSING SPERM CELLS AS PARASITES WAS AN ENORMOUS MIStake, but it was a blunder that stemmed from experimental mishap and not from ideological blindness. Along with his dilution experiments, Spallanzani had examined sperm cells with minute attention and concluded that they truly were animals. Semen samples from humans, horses, dogs, goats, bulls, sheep, fish, frogs, and salamanders all teemed with animalcules that looked and behaved just as you would expect from tiny, living organisms—they swam along purposefully, powered by their wriggling tails; they maneuvered past obstacles; they slowed down (and finally died) as they were cooled and moved about vigorously (and finally died) as they were heated, just like microorganisms that lived in pond water.

These were bona fide animals, then, and why would anyone expect animals of one species to have anything to do with sex and reproduction in the unrelated species where they happened to make their home?

Here we come to one of the most tantalizing near misses of the whole sex and babies saga. Although Spallanzani did not believe that sperm cells had any role in fertilization, he *did* wonder how these parasites propagated themselves. He offered a guess. Perhaps the tiny animals were passed along in semen during sex. If semen met an egg and fertilized it, the animalcules would enter the egg and make their way to the embryonic tissues that were destined to develop into genitals. (Presumably they would die off if they did not land in a congenial, male home.) There they would bide their time, waiting out the years until their new host produced semen of his own.

As dedicated and skilled an observer as he was, Spallanzani might have undertaken to test this theory. We can imagine him gazing through his microscope at a glass dish containing egg and semen. For once, he would have focused his attention not on the semen itself but on the sperm cells within it. If all had gone well, he would have become an eyewitness for the ages, the first person in history to witness sperm meet egg. What would he have made of that gigantic triumph?

Strangely, he would probably *still* not have concluded that sperm cells had anything to do with fertilization. More likely, he would have celebrated his correct guess that spermatozoa were parasites that made their homes in embryos. "It is rather alarming to think," writes the medical historian Elizabeth Gasking, "that had Spallanzani really seen the penetration of the eggs by the spermatozoa, he would have regarded it as a confirmation of this hypothesis." Seeing is believing, the saying goes, but the reverse can also hold. Sometimes believing is seeing.

CAUTIOUS AS ALWAYS, SPALLANZANI HAD ALSO LOOKED AT THE parasite question from another angle. These observations, too, conspired to mislead him. In experiments with toads, he wrote, he had almost always found their semen "very full of spermatic worms." Twice, though, he had been surprised to find no sperm cells at all. He had dabbed some of that sperm-free semen on eggs and found that they developed into tadpoles and then into toads. In follow-up experiments with frogs, he again found semen samples that contained no "worms" but that nonetheless proved capable of fertilizing eggs. Here was more proof that worms were beside the point. Spallanzani underlined his conclusion. "My long experience in the world of microscopical animals, whether belonging to man or animals, will, I hope, vouch for me that I was not deceived in this delicate investigation."

And yet he *was* deceived. The only plausible explanation is that Spallanzani, scrupulously careful though he was, had overlooked a sperm cell or two. Then, perhaps, he responded the way most of us do when experience seems to confirm our beliefs. He nodded contentedly and moved on to other questions.

The damage was substantial. So influential was Spallanzani, one historian of medicine remarks, that another full century would pass before scientists accepted the idea that spermatozoa played a role in fertilization.

Almost as remarkable as Spallanzani's neglect of sperm cells was this: despite all he had discovered about how semen worked, he gave no thought to the notion that perhaps semen and egg deserved equal billing in the sex and conception drama. Instead, like his fellow ovists, he focused nearly all his attention on his beloved eggs. He fervently believed that the egg was the headline act, and semen merely a bit player whose role was limited to providing the chemical nudge that set the fertilization process in motion. (It did so by jump-starting the embryo's heart.)

This was another profound mistake, but it, too, arose from painstaking experiments and careful deduction. After he had finished his work on the properties of semen, Spallanzani had turned his attention to eggs. How did fertilized eggs differ from unfertilized ones? After long hours with scalpel and microscope and countless rounds of poking, probing, comparing, and dissecting, the truth was plain. Unfertilized and fertilized eggs did not "differ in the least." From that starting point, the argument rolled out with the inevitability of a proposition in logic.

Spallanzani spelled it out: If unfertilized eggs were identical to fertilized eggs, and if fertilized eggs developed into tadpoles and then into frogs, then it followed beyond the possibility of dispute that the tadpoles were already present in the *un*fertilized eggs. The tadpoles had been there all along! The semen's contribution was essentially to raise the curtain and shout, *Voilà!*

This was, in the 1700s and well into the 1800s, an entirely mainstream view. When Leeuwenhoek died, back in 1723, the spermists had lost their last great advocate. Ever since, the egg had reigned supreme. (This was not seen as something to do with frogs only, but as a general fact of nature.) From a historical vantage point, this is extremely odd. The belief that had prevailed for millennia—that men planted seeds and women nurtured them—had nicely bolstered the self-regard of male scientists. But that theory had given way to a picture of males as all but irrelevant, even though the very scientists proposing the new view still worked in an almost exclusively males-only club and still lived in an almost entirely male-run world.

Somehow no one seemed to notice. This curious state of affairs prevailed for almost a century and a half, from Leeuwenhoek to Darwin. Throughout the plant and animal kingdom, the historian John Farley observes, "Reproduction was a uniquely female occupation in which the role of the male was very limited or even entirely unnecessary." Babies were women's work. So was making them.

That view was in large part due to Spallanzani. He was a brilliant thinker and an unsurpassed experimenter. He had sped ahead of all his rivals. His proof of semen's key part in fertilization and his demolition of "seminal aura" were giant advances. But he had missed a colossal chance. He might have looked harder at the swimming animalcules under his microscope and wondered if they could be up to something. He might have thought again about the dog he had artificially inseminated and wondered why it was, if fertilization was a story where the female played the only lead, that the puppies looked like both parents and not just their mother. He might have examined a drop of semen from his filter papers under his microscope.

He might have. But he didn't.

TWENTY-ONE

THE CRAZE OF
THE CENTURY

LAZZARO SPALLANZANI AND HIS PEERS TOOK AS THEIR MISSION sorting out the properties of life's building blocks. But life was more than an arrangement of constituent parts. Something more was needed. Some fluid, some force, some spark had to jolt those parts to life.

Cue electricity! In the 1700s, scientists believed they had finally found, in electricity, the "vital force" that explained the gulf between life and nonlife. Which was why, in a Paris courtyard in 1772, a physicist named Joseph-Aignan Sigaud had arranged sixty nervous volunteers into a human chain, each man holding his neighbors' hands and waiting anxiously for Sigaud to finish charging his electrical generator.

Demonstrations like this one were wildly popular. At a signal, Sigaud shocked the first man in line, who jumped in pain. As did the next and the next and the next, and so on, while the crowd whooped in delight. (The charge was strong enough to startle but not to harm.) But wait! Something had gone wrong. The shock had passed through the first six men but then stopped.

No one could think why. Sigaud tried again. Once more, the shocks and the yelps stopped with man number six. Everyone had a theory about what could be wrong with him. In the commotion, Sigaud didn't think to try the experiment without number six's participation. Soon the rumors converged on a single theme—the problem with the unfortunate Monsieur Six was that he lacked the animal vitality, the sexual vigor, that ordinary men possessed. In Sigaud's words, "the young man in question was not endowed with everything that constitutes the distinctive character of a man." In time, the great mystery reached the ears of the Duke of Chartres, who proposed a test. It so happened that among the duke's musicians were three castrati. Would these men be susceptible to electric shocks?

While the duke and a bevy of scientists looked on, Sigaud assigned places to some twenty men, including the three musicians. He sent a shock down the line. Everyone jumped, the musicians as high as the others, and everyone passed the charge to his neighbors. Baffled, Sigaud sent his volunteers on their way. Months later, he performed yet another demonstration. This time the experiment failed, with the electric charge once again making it only partway through the line of volunteers.

Sigaud repeated the experiment twice more. Two more failures, and each time the chain ended with the same man. (This was not Monsieur Six.) Sigaud examined him head to toe. Finally the light dawned—the last man to feel a shock had been standing on wet ground; when the charge reached him, it traveled down his legs and into the moist earth (as if he were a human lightning rod) rather than into the body of the next man in line. The mystery was a matter of puddles, not sex.

That electricity should have anything to do with genitals and castrati might not occur to anyone today. In the late 1700s, nothing could have been more plausible. Electricity was, in one historian's words, "the craze of the century." Whenever a spectacle was called for, someone would crank a handle and thrill a crowd with sparks and shocks. Whenever a mystery was invoked, electricity was the first suspect named. This was especially true if the mystery had to do with sex, energy, and life.

In the minds of eighteenth-century scientists, Spallanzani's painstaking experiments on semen and eggs and the electrical extravaganzas of physicists like Sigaud fit neatly together. (It was Sigaud's mentor who had suggested to Spallanzani that he put his frogs in boxers.) Electricity had to be a key part of the sex and conception riddle, scientists in the late 1700s felt, because *some* force plainly was, and all the traditional possibilities had been ruled out. For countless ages, it had been enough to invoke magic or spirit or soul. In the new scientific era, such notions felt outmoded, akin to explaining volcanic eruptions as temper tantrums of the gods. Electricity offered a newer, better answer.

It would turn out, to peek ahead for a moment, that the focus on electricity was an inspired guess. Humans and other animals *are* electrical machines, as everyone who has ever watched a heart monitor knows. Filling in the details in that story would prove immensely difficult. Scientists would not truly understand the electrochemical underpinnings of life until the 1970s. But Spallanzani and Sigaud and the other seventeenth-century pioneers had made a brilliant start.

They believed, mistakenly, that they had done far more than that. In electricity, scientists in the late 1700s decided, they had found the answer to *two* age-old questions. The first had to do directly with sex and conception. From Aristotle on, biologists who took on the mystery of babies had devoted most of their energy to squabbling over which bits of the body did what. Larger questions, about just where in the picture *life* came in, they did their best to duck.

That narrowness of vision was understandable—everyone prefers approachable questions to baffling ones—but also unfortunate. When it came to the biology of sex, scientists found themselves stuck. It was as if they had set out to study flight but ended up focusing all their attention on taxidermy and arranging stuffed birds in attractive settings, and none on asking how actual, living birds manage to lift off the ground and soar into the sky.

They had studied human and animal anatomy and scrutinized embryos and tissues and all the body's parts and products, but they had not been able to envision a force that would spark those odds and ends to life. And then electricity came along and seemed, at last, to offer an answer.

Here was a compact, thrilling story: men and women provided the physical bits that babies were made from, and electricity animated those dribs and drabs. And that was only the first of electricity's two crucial roles. In the first place, electricity provided the spark that zapped a body to life. Then, over the course of a lifetime, electricity provided the energy that kept those bodies running.

The "life force" that had been sought since humankind's earliest days had at last been found. Or so eighteenth-century scientists fervently hoped.

I N ANCIENT TIMES, NO ONE HAD REGARDED ELECTRICITY AS MORE than a curiosity. Through the millennia when no one understood lightning or knew about batteries or generators, electricity seemed too weak and fleeting to merit much scrutiny. *Have you ever noticed how sometimes on a dry day your hair stands up when you brush it?* Heat had seemed more promising. Perhaps living animals burn with a slow fire? Humans are warm, after all, and early thinkers had often pictured the heart as a kind of hearth. But wait! Snakes and frogs are alive, and everyone knows they're cold to the touch.

Through the ages, this had been the pattern—a big question, a bright idea, and then, almost at once, frustration and dismay. But the hunt for the life force was no mere puzzle. Here was a mystery that carried enormous emotional weight. A gulf separated a living, vibrant body from the same body, identical but so unaccountably different, when life had vanished. Fascinated and appalled, poets and philosophers had explored the theme for eons. Shakespeare confronted it repeatedly. Audiences wept with Lear as he held Cordelia's body in his arms and howled in grief: "She's gone for ever! / I know when one is dead, and when one lives; / She's dead as earth." Everyone knew that difference. No one knew how to account for it.*

Now scientists would take their turn. They had begun talking vaguely about electricity in the early decades of the 1700s. The idea was that all living beings—not just a few bizarre sea creatures known since ancient times—contained electricity, and that the force that had traditionally been known by such names as "animal spirits" was in fact "electrical fluid." Then came the discovery, in the 1740s, that electricity could be generated and manipulated and stored in large, dangerous quantities. What had long been merely an amusement was now transformed into a potent, if still mysterious, natural force.

As a bonus, probing that mystery made marvelous theater. Scientists and showmen performed for large and eager audiences. (Often the scientists *were* the showmen.) This was new; gravity had long been recognized as a mighty force of nature, but it did not rate high as a source of entertainment. *Pick up rock, drop, repeat.* Electricity, on the other hand, provided endless opportunities for creative mayhem. Traveling lecturers set off foot-long sparks, set oil afire, ignited gunpowder, melted metal, electrocuted animals.

So thrilling were these performances that they displaced ordinary entertainments. Bold men would volunteer to touch a crackling orb, and audiences would gasp and hoot when the victim jumped. In the most elegant homes, the *Gentleman's Magazine* reported in 1745, "electricity took place of quadrille." Who had time for dancing? One of the most popular demonstrations, performed across Europe, was known as the "electric kiss." A young woman selected from the audience would perch atop an insulated cushion or don a pair of glass slippers. The electric virtuoso would charge her body and then call for gentlemen volunteers, whose task was to give the lady a kiss. "Alas, as they tried to approach her lips a strong spark would discourage any

---

*In *Hamlet,* probably written a few years before *King Lear,* Shakespeare had put a more sardonic twist on the impossible-to-fathom fact that we are dust and return to dust: "Imperious Caesar, dead and turned to clay / Might stop a hole to keep the cold away. / Oh, that that earth, which kept the world in awe, / Should patch a wall to expel the winter's flaw!"

attempt," one historian writes, "while exhilarating the lady and the rest of the audience."

Performances were part educational lecture, part magic show. Electricity was always described as "wonderful," in the sense of "wondrous." To contemplate an invisible force that hurtled across the sky in lightning bolts, and made hair stand on end, and gave fish the power to stun anyone who touched them, was to gasp in awe at nature's bounty.

In England, France, Italy, even Poland, aristocrats sat next to shopkeepers at electrical shows, and the deeply learned and the merely curious squeezed against one another. In Germany dukes and duchesses "honored [the local electrical savants] with their presence, and their astonishment." In England, King George III proudly amassed a huge collection of scientific instruments, with electrical gizmos prominent among them. In France, Louis XV ordered up scientific demonstrations in the Hall of Mirrors at Versailles. In Austria, Emperor Joseph II entertained his guests by bringing them to scientific lectures.

Better than any lecture was the sight of a long line of men bracing themselves for a shock. (Sigaud and his castrati were latecomers to what was, by their day, a familiar entertainment.) This was not quite ten lords a-leaping, but it came close. A French abbot named Jean-Antoine Nollet, who was in charge of a monastery in Paris and a science buff, was one of the first to explore the possibilities. On an otherwise ordinary day in 1746, Nollet ordered two hundred of his monks to form an enormous circle, and then he distributed long, iron rods among them. Each man was to stand facing the center of the circle, his right hand clutching one iron rod and his left clutching another.

When all two hundred monks were linked together, the abbot connected the two ends of the circle to a Leyden jar, a powerful source of electricity. Two hundred startled monks leapt, bewildered and in pain, into the air. "It is singular to see the multitude of different gestures," one observer wrote happily, and he noted, too, "the instantaneous exclamations of those surprised by the shock."

Word quickly reached Versailles. Louis XV ordered a repeat performance, this time with 180 soldiers holding hands. Soon Europe was dotted with whole regiments of unfortunate soldiers leaping, against their will, into the air. In England, the record was 1,800 tingling, tormented soldiers taking one simultaneous bound.

B Y THE MID-1700S, SCIENCE BEGAN TO NUDGE SHOW BUSINESS to one side. Ben Franklin played a key role in the shift, though his electrical career nearly ended in its earliest days. On a December evening in 1750 Franklin prepared to electrocute a turkey for a Christmas feast. (His plan was to wire the bird to a kind of primitive car battery and then roast it.) He had managed the trick before, but this time, as he told the story to his brother, "I inadvertently took the whole [shock] through my own arms and body." Franklin's dinner guests reeled in fright from a flash of light and a crack that sounded like a pistol firing, but Franklin himself missed all the excitement. "My senses being instantly gone, I neither saw the one nor heard the other." He came to, eventually, when the "violent, quick shaking" of his body gave way to mere numbness.

Two years later, Franklin ventured into a thunderstorm with a kite. We think of the story as a scene from folklore, like George Washington and the cherry tree, but it really happened, and it demonstrated that the lightning bolt that flashed through the sky and the spark from a dry finger reaching toward a doorknob were identical in nature.*

The old belief had been that lightning was a form of divine fire, which explained why church steeples were hit so often during storms. (Lightning was sometimes called "heaven's artillery.") One response was to send bell ringers up the steeples during electrical storms, on the theory that the sound of the bells would ward off God's wrath. This proved irrelevant to the lightning and calamitous to the bell ringers.†

---

*Though Franklin did not know it, a French scientist had beaten him to the punch by drawing lightning from a storm one month earlier. Franklin had proposed the experiment in print, and the Frenchman, Thomas-François Dalibard, had followed up at once.

†Many religious believers opposed lightning rods, on the grounds that it was sinful to try to thwart God's plans. Nollet, the French abbot, warned that it was "as impious to ward off God's lightnings as for a child to resist the chastening rods of the father."

These early experiments were profoundly dangerous. The first person to die by electrical mishap, in 1753, was a German physicist named Georg Richmann, who was trying to follow Ben Franklin's instructions for building a lightning rod. A ball of lightning raced through the rod and into the room where Richmann was standing, knocking the door off its hinges, ripping open one of Richmann's shoes, jumping to his forehead, and killing him instantly. Historians cite Richmann as the first experimenter to die of electric shock and, simultaneously, the first person to observe ball lightning.*

All the early "electricians," as they called themselves, knocked themselves silly. "The first time I experienced it," one English physicist wrote, it felt "as though my arm were struck off at my shoulder, elbow, and wrist, and both my legs at the knees and behind the ankles." As soon as he recovered, he rigged up a room with a hidden battery and wires concealed in the carpet, in order to provide unsuspecting friends the same experience. Another electrician, who had shocked himself so severely that blood spurted from his nose, decided that it was too dangerous to continue experimenting on himself. He recruited his wife instead. Soon he reported that an electric charge had knocked her to the ground and left her temporarily unable to walk.

Those injuries paled next to the torments that one German physicist inflicted on himself. Johann Ritter was a well-regarded experimentalist who, in the name of science, systematically zapped himself with large currents of electricity. First he sent electricity coursing through his entire body, and then he worked through each sense in turn. "Instead of one of the two hands," he wrote, "one brings an eye, an ear, the nose, the tongue, or another part of the body into the closed circuit."

Ritter's notion was that sights, sounds, smells—in fact *all* the sense impressions of the body—originated in electric signals. By way of proof, he found that hooking up electrodes to various body parts sent blue and red lights flashing before his eyes, tones ringing in his ears, cold and heat coursing through his fingers (and sneezing fits racking

---

*Joseph Priestley, one of the eighteenth century's most admired scientists, remarked that "it is not given to every electrician to die in so glorious a manner as the justly envied Richmann."

his whole body). Convinced that he had still more secrets to unearth, he proceeded to attach electrodes to the "organs of reproduction" and "the organs of evacuation," as he put it, as well as to "other choice parts of the body." By age thirty-three he was dead. The cause was unspecified, but we can venture a guess.

IN ENGLAND, IN THE LATE 1700S, ELECTRICAL EXPERIMENTATION had followed an altogether cheerier course. For a few years around 1780, London's most fashionable lords and ladies were agog with excitement about the sex theories of a dashing, impossibly articulate, not-quite-doctor named James Graham. A Scottish-born medical school dropout, Graham had spent time in America, where he met Benjamin Franklin and concocted his own theory of medical electricity.

In London Graham opened a Temple of Health that was part theater, part lecture hall. Visitors passed through a series of rooms that proclaimed a link between sex and electricity. Graham, who believed that whatever was worth doing was worth overdoing, made sure no one could miss the point. The Temple's first room featured an enormous metal cylinder, about eleven feet long and one foot thick, resting atop two half-globes and "blazing with electrical fire." That fire in turn set an enormous golden dragon to crackling and sparking. From the dragon, the charge passed to a ten-foot-tall (and insulated) throne. Here visitors perched in splendor and absorbed the rejuvenating rays of celestial fire.

At the Temple's heart was the most exciting, most celebrated object in Graham's collection. This was the Celestial Bed, a guaranteed cure for infertility and impotence. (An inscription on the headboard proclaimed, "Be fruitful, multiply, and replenish the earth.") The bed stretched twelve feet long and nine feet wide and boasted a ceiling-mounted mirror, music from a pipe organ, candles, flowers, and spice-scented breezes, all this accompanied by "the exhilarating force of electrical fire." For the stupendous fee of £50 a night—in the 1780s a workman might earn £50 in a *year*—Graham promised "any gentleman and his lady" who made use of the celestial bed that they would achieve "immediate conception," not to mention "superior ecstasy . . . never before thought of in this world."

The key, Graham told his hordes of eager visitors, was electricity. "Even the venereal act itself," he explained, "is in fact no other than an electrical operation!" This was not humbug, Graham explained, but established scientific truth. "In the first place . . . there is the necessary friction or excitation of the animal electrical tube or cylinder, for the accumulation, or mustering up of the balmy fire of life! This is what electricians call the charging of the vital jar. Then follows the discharging, or passage of that balmy, luminous, active principle, from the *plus* male to the *minus* female."

TWENTY-TWO

"I SAW THE DULL YELLOW EYE OF THE CREATURE OPEN"

F ROM ITALY CAME A MORE SOBER AND MORE IMPORTANT, though perhaps equally odd, attempt to demonstrate that "animal electricity" was the long-sought life force. A thoughtful, publicity-shy anatomy professor from Bologna had carried out a series of startling experiments. Luigi Galvani's claim was that electricity coursed through *every* animal, not just oddities like the sting ray.* Galvani conducted his experiments in the 1780s. They have been hailed as classics ever since. He worked with frog legs, severed from the body so that a large nerve lay exposed and dangling. What Galvani found was that electric signals could make those cut-off legs twitch and kick as vigorously as they had in life.

He had made his first frog observation "by chance," he wrote, though this was a characteristically modest way to downplay his flash of in-

---

*The ancient Greeks and Romans had known about sting rays (although not about electric eels, which are found in South American rivers). The physician to one Roman emperor recommended standing on a live sting ray as a cure for gout, because the shock numbed the foot and leg all the way up to the knee.

sight. (The reason that Galvani had frogs available in the first place, according to his earliest biographer, was that he planned to make a frog broth for his beloved but frail wife, Lucia.) One of Galvani's assistants happened to be working with an electric generator; independently, a second assistant was preparing a set of severed frog legs for study. No wire—no physical connection of any sort—linked generator and frog. The generator threw off a spark. At the same instant, the frog's legs gave a forceful kick!

In follow-up experiments, Galvani pinned up frog legs outdoors on a stormy day and saw that when lightning flashed, the legs kicked. That was fine, and just what Ben Franklin would have predicted. The surprise came when Galvani, who was notably careful and thorough, repeated the experiment on a sunny day. He poked brass hooks through frog legs and hung the legs on an iron railing. They twitched! By 1791, Galvani had sorted it out. Not only did animals *respond* to electric signals; they also *produced* electricity of their own. Here was the long-sought vital force, identified beyond a reasonable doubt.

The scientific world gaped in awe, as staggered by Galvani's discoveries, one historian remarked, as Europe had been by the French Revolution. But Galvani would soon be challenged. An eminent Italian physicist, Alexander Volta, had been thrilled when he first learned of Galvani's work. In time he changed his mind.

Volta made a formidable rival. Weighed down with scientific prizes and honors, he was combative, self-assured, and in happy possession of, in his own words, "a genius for electricity." In connections and temperament, though not in talent, Galvani was overmatched.

The clue that Volta jumped on was the twitching on a sunny day. As Volta explained—and as scientists now agree—Galvani had unknowingly created an electric current that flowed between the iron railing and the brass hooks. The presence of two different metals was the key; the frog was beside the point.

To test his theory, Volta skipped the frogs altogether and instead arranged a tall stack of alternating silver and zinc discs. In between each disc he inserted a piece of cardboard soaked in salt water. Then he wired up this metal and cardboard tower. "Volta demonstrated that an electric current flowed when the top and bottom of the pile were connected," one modern scientist writes. "He had invented the first electric battery." The frog was, so to speak, a red herring.

Volta had won the day, or so the world concluded. A clash that pitted biology (and Galvani) versus physics (and Volta) was not a fair fight. In 1800 biology was still struggling for respect and tainted by its links to medicine, which reeked of quackery. Physics basked in a glow cast long before by Galileo and Newton and their fellow titans. Biology was a scruffy upstart, physics a haughty ruler, and that contest was no contest.

In hindsight we can see that both Galvani and Volta had grasped part of the truth. It would take nearly two centuries for the details to emerge, but it would finally become clear that, in the body, biology and physics work in tandem. Bodies are made of cells—Galvani and his contemporaries had no idea—and each of those microscopic cells is a kind of chemically powered battery. As a result, every living creature is a sound-and-light show, with Galvani and Volta, the one-time foes, as coproducers. Every thought and sensation we experience is an electrochemical extravaganza. When Walt Whitman exclaimed, in 1855, "I sing the body electric," he spoke truer than he knew.

GALVANI DIED AT AGE SIXTY-ONE, IN 1798, HIS BATTLE WITH VOLTA still unresolved. (The world "laughs at me," Galvani supposedly lamented, "calling me 'the Frog's dancing-master,' but I know that I have discovered one of the greatest Forces in nature.") In London, in 1803, his nephew Giovanni Aldini prepared a demonstration that his earnest, quiet uncle would never have permitted. Aldini was a showman with a taste for controversy and the macabre. Galvani had carried out innumerable meticulous experiments on frogs. Aldini's style ran more to zapping electricity through the body of a freshly hanged murderer, while a crowd gasped.

After Galvani's death, Aldini took over the family business, preaching the electric gospel across Europe. His credentials were impeccable. He was a professor of experimental physics at the University of Bologna, and he had begun working with Galvani on his frog experiments, in a makeshift laboratory in Galvani's home, as soon as he had graduated from college. Once on his own, he moved from experimenting on cold-blooded animals like frogs to warm-blooded ones, starting with birds, sheep, and oxen. Then came human corpses.

In the winter of 1802, near the Palace of Justice in Bologna, Aldini fired up an electric battery. As spectators looked on, Aldini wrote, "the first of the decapitated criminals was brought to the room I had chosen." The men had been executed an hour before. Aldini attached wires from his battery to various points on the body, which twitched in much the way a frog did. Then he connected wires to the two ears of a severed head. The muscles of the face contorted in "the most hideous grimaces," and the dead man's eyes blinked open and shut.

Aldini continued on to England. He lectured on animal electricity in Oxford and London and sent electric charges coursing through corpses in the anatomy theaters of London's leading hospitals. Dukes, doctors, and, once, the Prince of Wales looked on in fascination and horror.

Aldini pulled off his most spectacular coup in January 1803. A London man named George Foster had been convicted of murdering his wife and infant daughter. Too weak or too despondent to walk, Foster was dragged up the stairs to the gallows and then, the newspapers reported, "launched into eternity." The body (with head intact) was cut down and rushed to the Royal College of Surgeons, where Aldini and an eager crowd waited.

Aldini connected one metal rod to the corpse's mouth and another to an ear. Foster's jaw quivered, and his facial muscles clenched and relaxed. He opened one eye. Aldini moved his metal contacts. Foster lifted one hand and made a fist. His legs shook and kicked. Spectators

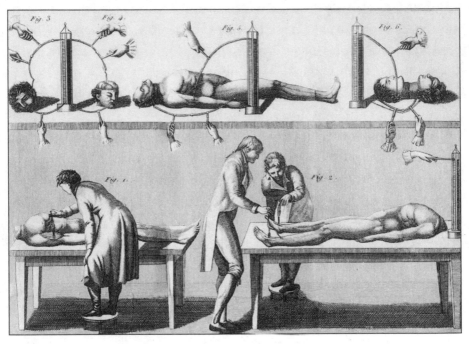

FIGURE 22.1. Giovanni Aldini attempts to animate the head of a corpse (above left) and a corpse without a head (bottom).

gasped that Foster had come back to life, and one man fled the room and, according to newspaper reports, promptly died of fright.

For nearly twenty years more, doctors would continue sending jolts of electricity through the bodies of newly executed criminals. (In Scotland, in 1818, one dead man kicked out a leg so violently that he nearly knocked over the surgeon's assistant; the corpse's chest rose and fell, as if he were breathing, and his fingers "moved as nimbly as those of a violinist." Spectators vomited and fainted.)

Aldini and the others had come near an important truth—they had grasped that electricity is the force that powers living organisms, though they had not understood how that happens—but they did not inspire further generations of scientists to follow their lead with experiments of their own.

Instead, Aldini inspired Mary Shelley to create one of literature's enduring horror stories.

June 1816, near Lake Geneva Summer has been cold and rainy, and the poet Byron and three visiting friends have found themselves trapped indoors yet again. Byron is only twenty-eight but already an international celebrity. Four years have passed since he "awoke one morning and found myself famous." He has been called "mad, bad, and dangerous to know," and he is working mightily to live up to the description. Caught up in a swirl of sexual scandals and chased by bill collectors, he has left England barely ahead of his pursuers. He left in style. His coach was an exact copy of Napoleon's; he was attended by a valet, a footman, and a personal physician; his entourage included a peacock, a monkey, and a dog.

No one yet knows Percy Shelley. In 1816 the future poet is simply a strange and bookish young man who was booted out of Oxford a few years before for advocating atheism. Shelley's lover is a brilliant young woman, Mary Godwin. Only eighteen, she has lived with Shelley since they eloped when she was sixteen. She has already given birth to two children—the first died at two weeks of age—and she is pregnant again. Standing at the group's edge trying to get a word in is John Polidori—"poor Polidori," Mary Godwin calls him—a much-derided doctor and aspiring writer supposedly tending to Byron's health. Polidori attended the University of Edinburgh during the heyday of its body-stealing, grave-robbing era.

They talk deep into the night, evening after evening. All four are fascinated with the latest scientific developments, especially those having to do with electricity and life's "vital forces." Percy Shelley's obsession runs deepest of all. Even in childhood, his hands had been perpetually spotted and stained with chemicals, his clothes dotted with holes where acids had splashed on him. The more dangerous an experiment, the more appealing. Growing up, Shelley enlisted his sisters as his involuntary test subjects, and one later recalled how "my heart would sink with fear at his approach."

The sight of wires and other bits of electrical apparatus, the better to generate shocks, was especially bad news. "We were placed hand in hand round the nursery table to be electrified," Helen Shelley recalled

FIGURE 22.2. Drawing of the monster from the original edition of *Frankenstein*.

miserably. At Oxford, Shelley's room overflowed with vials and beakers for his forays into chemistry, and visitors had to navigate a path between microscope, telescope, air pump, and an assortment of electrical contrivances. At age eighteen, he wrote a letter proclaiming that "man is no more than electrified clay."

One gloomy night by Lake Geneva, Byron issues a pronouncement: "We will each write a ghost story." They compare notes every morning. Mary has nothing. Come evening, Byron and Shelley rattle on again, while Mary—"devout but nearly silent," as she recalls later—lets the words wash over her. The men discuss the strange experiments of Erasmus Darwin.

He is nearly forgotten today, but Erasmus was a prominent physician, an early advocate of evolution (and Charles's grandfather), and a lively poet famous for his bold ideas and his good cheer. (He grew so fat from a lifetime of lavish meals that he had a semicircle cut from his dining room table to accommodate his belly.) Byron and Shelley tell how, as Mary recounts it later, Erasmus had "preserved a piece of vermicelli in a glass case, till by some extraordinary means it began to move with voluntary motion." The poets gasp. If a strand of *pasta* could come to life, what else might be possible? "Perhaps a corpse would be re-animated," Mary writes. "Galvanism had given token of such things: perhaps the component parts of a creature might be manufactured, brought together, and endued with vital warmth." Thus was *Frankenstein* born.*

In hindsight, it seems that Byron and Shelley had not quite understood. Erasmus had written not of "vermicelli" but of "vorticellae," microscopic creatures found in pond water. No matter. Mary Shelley grabbed her pen. "It was on a dreary night of November that I beheld the accomplishment of my toils. . . . I saw the dull yellow eye of the creature open; it breathed hard, and a convulsive motion agitated its limbs."

---

*Polidori's contribution to the ghost story contest was *The Vampyre*, the first novel built around the romantic, murderous adventures of a blood-sucking aristocrat.

TWENTY-THREE

THE NOSE OF
THE SPHINX

B Y THE EARLY DECADES OF THE 1800S, THE AGE OF MARY Shelley and Frankenstein, biologists found themselves stymied. Life seemed fated to remain a dark mystery. Galvani thought he had answered a crucial question—*What is the source of the power that propels living creatures?*—but Volta had shot him down. Back to square one.

In the meantime, the question of "vital force" remained crucial and unresolved. The blindest man could see that living creatures have some source of power that lets them carry on moving and digesting and growing. Moreover, those batteries or engines or whatever they might be work on their own, automatically, without any outsider intervening to throw a switch or turn a key. (Life would be far less mysterious if babies looked like wind-up dolls, with keys sticking out of their back, and parents dutifully wound them up each morning.) If it wasn't electricity that provided that power, what did?

The temptation was to say, "Food." But that was no answer, everyone saw with chagrin, because that supposed explanation led immediately to an equally baffling question, "And how does *that* work?"

You could cram bits of food inside a doll forever, after all, and the doll would never whir into motion.

Past eras had concluded that the question was out of reach and moved on, but a scientific age demanded some sort of answer. Three centuries had passed since Michelangelo had depicted God passing a divine spark to Adam through an outstretched hand. The image was as glorious and uplifting as ever, but now it served more as an emblem of the mystery than an answer to it.

Whatever the vital force was, it was evidently in short supply. We humans see life all around us, because we cannot help putting ourselves at the center of every picture. But life is extraordinarily rare. If all the world's a stage, it is an almost entirely empty stage in an empty auditorium. The present-day physicist Alan Lightman has tried to put numbers to that barrenness. "Only about one millionth of one billionth of one percent of the material of the visible universe," he estimates, "exists in living form."

The urgent task, as the nineteenth century saw it, was finding what distinguished those few precious bits. What set living organisms apart from nearly everything else? The quest to find out took two distinctly different paths. The first followed directly from the work of Spallanzani and Leeuwenhoek and their anatomist forebears. The idea was to look ever more closely at egg and sperm to try to sort out the mechanics of fertilization. *To learn where life comes from, learn where babies come from.* The other path took more of a bird's-eye view. The aim here was not to sort out the anatomical details of sperm and egg but to tackle a far broader question: *What does it mean to be alive?*

The curator of the botany collection at the British Museum believed staunchly in this second approach. But not for him any of this foolishness involving batteries and shocks and jumping soldiers. Robert Brown focused instead on his beloved plants. In 1827, he set himself a curious mission.

Brown had spent his career staring at plants under a microscope. He would tackle the riddle of the life force as he had taken on countless challenges, by staring and thinking. Quite likely he would not see anything directly. But perhaps he could learn about his quarry by examining its effects, as a meteorologist might study the wind by looking at trees exposed on a mountainside. Brown began by taking pollen grains and sprinkling them in a drop of water. (His notion was that pollen, from the male parts of a plant, would be more active than female bits, which would no doubt sit dully and passively in place.)

To Brown's delight, the pollen grains never settled down quietly. They jiggled, and they kept jiggling. This was surely the vital force in action. With the entire British Museum collection to draw from, Brown carried out a series of follow-up experiments. He ground up pollen grains from a variety of recently gathered plants and from plants that had been dead for a century. The result was always the same— the pollen grains kept up a perpetual dance. This was almost *too* good. The life force seemed to persist beyond death! Now Brown ground up the female parts of plants. Bizarrely, those grains danced, too.

Brown retreated. The life force evidently had to do with sexuality in general and not solely with maleness, as he had expected. Then came a fortunate accident. Careful though he was, Brown happened to contaminate one of his experiments. Bits of a crushed leaf—neither male nor female—fell into a water drop. Brown took a close look. The leaf particles jiggled, too!

Brown retreated further still. Inside every bit of every organism, he proposed, whether it was now living or had lived long ago, lay tiny, hidden particles steeped in "vital force." Over the course of the next year, Brown tested this new theory thoroughly and methodically. He ground up plants and vegetables of all sorts, and then bits of animal tissue, then flecks of coal (from prehistoric plants), and specks of petrified wood. They all jiggled.

With all those confirmations of his theory in hand, Brown might have declared victory. Instead, and admirably, he devised a test that might bring everything crashing down. He ground up bits of glass, which had never been alive. They danced as energetically as specks from a green and thriving plant! Brown tried pieces of metal; they jiggled. So did fragments of rock. As a grand finale, Brown took the deadest thing he could think of—a tiny speck from the nose of the Sphinx—and dropped some crushed particles into water. They jiggled!

Brown gave up. So did everyone else. The dancing movement he had identified was named "Brownian motion," but no one in the nineteenth century managed to make sense of it. Finally, in 1905, Albert Einstein explained what was going on.* At the time, scientists had not yet agreed on whether atoms and molecules were real, physical objects. Einstein argued that they were, and he pointed to Brownian motion as proof. Look at the way a speck of dust dances around on the surface of a glass of water, Einstein said. Its zigging and zagging tells us that it is being kicked this way and that by real but too-small-to-see particles within the water.

Einstein won the day. Back in 1828 Brown met a harsher reception. But he had, against his will, deeply undermined the theory that living objects contain an animating, intangible force that nonliving objects lack. More challenges would come, and in the same year of 1828.

CLOSE KIN TO THE THEORY OF VITAL FORCE WAS THE IDEA THAT living organisms and lifeless ones were made of different building blocks. (The term "organic chemistry"—this is the class that has tormented generations of premed students—reflects this now outmoded belief.) The claim wasn't that a real dog and a stuffed-animal dog differed in every single bit, just that *some* components of living creatures were different. Those magical parts could only be found in living animals, and never in the laboratory, presumably because their preparation required a dollop of vital force.

One much-cited example was urea, a substance found in urine and, as far as was known, nowhere else in the world. But in 1828 a German scientist named Friedrich Wöhler managed to produce urea from indisputably nonliving materials. "I can make urea without need of a

---

*This was Einstein's "Annus Mirabilis," the miraculous year when the twenty-six-year-old patent clerk published *four* revolutionary papers. One explained Brownian motion, another proposed that light comes in packets called quanta, a third unveiled special relativity, and a fourth announced that $E = mc^2$.

kidney or even an animal, be it man or dog," Wöhler wrote proudly to a friend.

Wöhler had delivered a body blow to the age-old belief that the mysteries of life lay beyond the reach of science, and biologists and chemists greeted his coup with jubilation and astonishment. Another blow followed soon after.

This latest assault was a redo of a classic experiment. Decades before Wöhler, at about the time of the French Revolution, the modern science of chemistry had taken its first baby steps. Antoine Lavoisier was one of the proud parents. Lavoisier was a genius, a nobleman at a time when that meant trouble, and a meticulous researcher. In a dazzling career, he had racked up countless breakthroughs. Lavoisier had been the first to explain the ancient mystery of fire. *What goes on when something bursts into flame?* And it was Lavoisier who had established the fundamental truth that, no matter how you burn or break or freeze or cook anything whatever, it weighs precisely as much afterward as it did at the beginning. You can transform matter, but nothing you do can conjure up something out of nothing (at least not anything that shows up on a scale) or make anything vanish.*

In one painstaking experiment, Lavoisier compared the heat output from a living animal with that from a piece of burning coal. He put a guinea pig in a container with ice-filled walls and then did the same with a lump of coal, and measured how much ice each one melted. (To make sure that he was not just putting ice out to melt in the sun, Lavoisier performed his experiment on a freezing-cold winter's day.) Then he put the guinea pig and the burning coal inside a bell jar and measured how much carbon dioxide they produced. The result was that breathing and burning produced about the same amount of heat for a given amount of carbon dioxide. Breathing was slow burning.

---

*Lavoisier's discoveries could not save him from the guillotine. He was executed in 1794, at age fifty, for his role in a tax-collecting scheme. A judge turned down an appeal to save his life. "The Republic needs neither scientists nor chemists; the course of justice cannot be delayed." So runs the famous quote, at any rate, though it may be apocryphal. But no one has ever challenged the judgment of Lavoisier's contemporary, the mathematician Joseph-Louis Lagrange: "It took them only an instant to cut off his head, but France may not produce another such head in a century."

This did not prove that a living animal and a black lump were the same, but it did show that, as different as they might be, they both obeyed the same scientific laws. If there was a "vital force" that set life apart from other things, it was measurable, not mystical. Lavoisier had shown, in effect, that if witches do fly through the night, they obey traffic signals as they go.

That had been back in the 1780s. Six decades later, in 1848, one of the nineteenth century's most eminent scientists, Hermann von Helmholtz, tightened the shackles on the life force even further. In the years since Lavoisier, scientists had found that his measurements were off a bit. Animals seemed to produce about 10 percent more heat than they should. All those scientists who believed that life could not be reduced to chemistry and physics cheered that discrepancy—*that unexpected 10 percent is the vital force at work!* But Helmholtz's new experiments showed where the error had crept in and dashed the vitalists' hopes.

This made for an odd impasse. Every time you tried to reach out and grab the life force in your hands, it vanished. Did that prove that the whole notion was misguided, a relic of earlier times like black magic and the evil eye? This was the view of hard-core mechanists like Helmholtz. But in the first half of the 1800s most scientists (and nearly all laymen) disagreed. Plainly *something* accounted for the difference between a living creature and the ground under its feet.

F. Scott Fitzgerald famously suggested that the test of a first-rate mind is the ability to believe two contrary ideas at once and still carry on. For nineteenth-century biologists, the challenge was almost the reverse. They had to proceed onward without losing heart while holding only *half* of a clear idea in mind. *Life has something distinctive about it, and we only wish we knew what that could be.*

Throughout the years that the *What does it mean to be alive?* camp had occupied themselves studying powdered nose-of-sphinx and watching guinea pigs shiver in icy containers, the *Where do babies come from?* faction had stuck close to the mysteries of sperm and egg.

They had found something remarkable.

TWENTY-FOUR

"THE GAME IS AFOOT"

IN THE 1820S, WHILE ALL EUROPE WAS HAPPILY SETTLING IN WITH *Frankenstein* for a night's spooky reading by the fire, two young biologists took on the riddle of life from a different angle. Fifty years had passed since Lazzaro Spallanzani had dressed his frogs in boxer shorts. In all those years scientists' understanding of conception had scarcely advanced.

In particular, biologists still echoed Spallanzani's two best-known claims. First, when it came to fertilization, semen and egg had to make physical contact. "Nearby" was not good enough, and "auras" and "emanations" were fictions. Second, spermatozoa were parasites that had nothing to do with fertilization. The first was correct, the second wildly wrong.

The explanation for the long lull was simple enough: the riddles of sex and development in particular, and the mystery of life in general, seemed so daunting that no one knew how to take them on. Across Europe, scientists had responded to that predicament in distinctive ways. In Germany, where the Romantic movement had taken hold, they veered toward the mystical, with grand-sounding theories about life's relentless impulse to strive ever upward. In England, scientists had virtually abandoned biology, which had been left to amateurs who puttered away with plants and pigeons. In France more than anywhere else, the experimental tradition had endured.

But no one was hopeful. The English physician and scholar Peter Roget* summed up the prevailing view. The mystery of where babies come from, he wrote in 1834, "surpasses the utmost powers of the human comprehension." Science could not offer "the least clue" toward unraveling "this dark and hopeless enigma."

A decade before, two young colleagues had in fact unearthed some considerable clues, although neither Roget nor anyone else at the time paid much heed. In three papers in 1824, Jean-Louis Prévost and Jean-Baptiste Dumas looked again at Spallanzani's work from so many years before. Dumas was French and still in his twenties, Prévost Swiss and a few years older. They began by studying semen from all sorts of animals. Whether they examined mammals, birds, or fish, they always found spermatozoa. And when they looked at infertile animals like mules and hinnies (the offspring of a male donkey and a female horse, or vice versa), they found no spermatozoa.

So far, this was simply endorsing Leeuwenhoek and Spallanzani. But now the two scientists began a series of experiments that moved beyond the work of their famous predecessors. First they put a batch of frog eggs into plain water and another batch into water with semen stirred into it. As expected, only the eggs in the water-with-semen developed. Then came the key. Spallanzani had dismissed spermatozoa as parasites. Was he correct that the spermatozoa played no role in fertilization?

Prévost and Dumas dried out a sample of semen, so that it contained no moving spermatozoa. Then they dissolved it in water. Would that semen still have the power to fertilize frog eggs? The answer was an emphatic *no*. Suppose you zapped a semen sample with electricity,

---

*Roget made no outstanding contributions to science, but he was absorbed, rapt, transfixed, and riveted with lists and order, and the great work of his life was compiling *Roget's Thesaurus*.

thus killing the spermatozoa. Was that semen potent? Again, *no*. If you had no swimming sperm cells, you had no fertilized eggs.

Next the two redid Spallanzani's filter paper experiments. After passing a semen sample through five layers of filter paper, it no longer contained any spermatozoa, and it could not fertilize frog eggs. But if you took the blob left on the paper and dissolved it in water, it did contain spermatozoa, and it did fertilize eggs.

This was not proof that spermatozoa were the male's crucial contribution to fertilization, but it surely pointed that way. Where you had spermatozoa, you had fertilization. Where you had none, you didn't.

Prévost and Dumas shifted their focus from sperm to egg, and from frogs to dogs and rabbits. Here, too, they came tantalizingly close to an enormous breakthrough. They almost, but not quite, identified a mammalian egg. No one had ever done so. Within what are now called the Graafian follicles in the ovaries, they found tiny, oval structures hiding under the surface of about-to-rupture follicles. They suspected, correctly, that these were eggs, but they could not demonstrate that this was so. (They would have had somehow to label these eggs and follow them on their journey to the womb.)

The world of science continued on its way, unperturbed. Spallanzani's reputation was too high, and the notion that spermatozoa were parasites too entrenched, for these new findings to make a splash. Worse still, to focus on spermatozoa was to endorse a doctrine that had been cast aside decades and decades before. What modern thinker wanted to return to Leeuwenhoek and his "animalcules"?

The bigger problem was that biologists in this era had gathered a giant heap of facts, but they had not begun to create a framework that made sense of them. Instead, they labored away like magpies collecting keys and rings and other shiny objects. Outsiders had mocked these diligent, misguided efforts for nearly a century. Biologists might "take pleasure in boring us with all the wonders of nature," one physician-turned-philosopher had complained in the mid-1700s, but they needed to move beyond "counting little bones in certain fishes" and "measuring how far a flea can jump."

Skeptics delighted in pointing out that physicists had moved beyond lists and catalogs long before, with their discovery of a handful of laws that applied universally. Rocks fell, and arrows fell, and the moon fell, and they all fell in precisely the same way. The Earth spun and so did skaters and wooden tops, and the identical rules applied to all of them. One all-embracing force pulled the planets toward the sun and a baby's rattle to the ground.

But biologists, and thinkers generally, had nearly given up hope of finding any such unifying laws for the living world. "Purpose" and "urges" and "drives" were the essence of life, and they could never be explained in purely physical terms. Physicists had a more manageable task. To ask, *What are stars for?* seemed silly. Stars weren't *for* anything. Neither were rainbows or rocks or any of the inanimate features of the world. They happened to turn up when the conditions were right.

When it came to biology, the picture looked completely different. To ask *What are eyes for?* seemed not merely sensible but essential. Eyes are for seeing, for finding one's way in a dangerous and complicated world. How could you possibly talk about eyes without starting there? Rocks just fall; they don't plummet through the air because they're in despair or trying to impress their friends. But nothing alive "just" happens. Even a dog scrounging through the trash or a cat chasing a mouse has a reason for its actions.

And the elaborate structures in the living world surely did not just happen to arise. How could you possibly talk about brains and bones and roots and flowers without asking how they had come to be and what they were for? Such questions were unavoidable, but they seemed doomed to go unsolved.

Biologists had set out eagerly in pursuit of an irresistible question— *How do living creatures move and reproduce?*—and they had found themselves stymied and helpless. In the pursuit of life, science could see only so far, even in the simplest cases. "There will never be a Newton for a blade of grass," Immanuel Kant declared, in 1790.

For decades, Kant seemed to have it right. Then the world shifted.

P RÉVOST AND DUMAS DESERVE A MEASURE OF THE CREDIT.
Though largely passed over, they had spurred a few scientists to
look anew at old questions. One was a young biologist named Karl Ernst
von Baer, born in Estonia but working at the University of Königsberg
in Germany. (By happenstance, this had been Kant's university.) In
1827, Baer became the first person to see a mammalian egg.

Baer was no Newton, but he did share the English scientist's
faith that straightforward experiments trumped high-flown talk. To
ask, *What is light?* was to invite endless prattle. The way forward,
Newton had demonstrated, was to hold a prism up to a beam of
sunlight and unweave the rainbow. Baer's question was, *What is
life?* He tackled it by searching inside a dog's ovaries, as Prévost and
Dumas had.

"When I observed the ovary," he wrote, " . . . I discovered a small
yellow spot in a little sac. Then I saw these same spots in several others, and indeed in most of them—always in just one little spot. How
strange, I thought, what could it be?"

"I opened one of these little sacs," Baer continued, "lifting it carefully with a knife onto a watchglass filled with water, and put it under
the microscope. I shrank back as if struck by lightning, for I clearly
saw a minuscule and well-developed yellow sphere of yolk. Before I
found courage to look at it a second time, I had to recover, since I
was afraid of having been deluded by a phantom. It seems odd that
a sight expected and so much longed for could frighten one when it
actually occurs."

Baer went on to study eggs of all sorts, from crayfish, birds, frogs,
lizards, snakes, and, notably, a host of mammals including rabbits,
pigs, cows, hedgehogs, and, especially, human beings. He drew a simple conclusion: "Every animal which is begotten by a sexual union
develops out of an egg."

This was an insight that had been a long time coming. And this
time—in contrast with Harvey's pronouncement in 1651 that "everything comes from the egg"—it was a conclusion based not on argument and analogy but on evidence and observation. It had taken so

long mainly because mammalian eggs were too tiny and well-hidden to find with the naked eye or with early microscopes.

But Baer had overcome psychological hurdles, too. Seventy-five years before Baer's lightning bolt, back in 1752, the great Swiss anatomist Albrecht von Haller had not only given up the egg hunt after a long, futile search but effectively scared away everyone else from hunting, too.

Haller had dissected forty ewes soon after they had mated, much as Harvey had dissected deer. Like Harvey, too, Haller found no signs of an egg or anything else in his cut-open animals. Finally, two weeks after the sheep had mated, Haller found a tiny embryo in the womb. His conclusion—which became the conventional wisdom taught to generations of young students, including Baer—was that the supposed egg that emerged from the ovaries was in truth a fluid that "curdled" in the womb and formed the embryo.

Where Haller had lost his way, few others dared venture. Baer did. He announced his findings in a paper with a grand but awkward title, "On the Genesis of the Egg of Mammals and of Man." The title was redundant, since humans were surely mammals, but Baer wanted to leave no room for doubt about what he had found.

His reward was meager, at least at first. The world responded to his breakthrough, Baer reported forlornly, with "deep silence." At a meeting of the Society of Naturalists, in Berlin in September 1828, no one mentioned his paper. Unwilling to raise the topic himself, Baer let it go. On the last day of the meeting, one scientist finally asked Baer a casual question.

In any case, Baer had seen only half the puzzle. He continued to insist—despite the work of Prévost and Dumas—that spermatozoa were parasites and irrelevant to the mystery of fertilization. (It was Baer who came up with the name "spermatozoa," meaning "animals of the semen.")

The breakthrough came a decade later, with the realization that spermatozoa were *not* animals but something quite different.

TWENTY-FIVE

CAUGHT!

O N AN OCTOBER EVENING IN 1837, A BIOLOGIST NAMED THEODOR Schwann and a lawyer-turned-botanist named Matthias Schleiden met over dinner to discuss their work. Both men were brilliant and high-strung. Schlciden, who suffered from depression, had changed careers after a failed suicide attempt. Schwann would suffer a religious crisis in 1838 and abandon his research career soon after. Friends since their student days in Berlin a decade before, the two chatted away excitedly.

Schleiden reported striking news: after endless hours staring at plants under a microscope, he had found the secret of their structure. No matter what species you looked at, plants were formed of cells—countless distinct units arranged just so. Schleiden had been following up observations by the botanist Robert Brown—this was the Brown of "Brownian motion" and the Sphinx's nose—who had noted that when he looked at orchids under the microscope, he found cell-like structures. More than that, each cell contained a round structure that Brown had dubbed a "nucleus."

Schwann did a double take. He had been studying animals, not plants, but he had seen dark spots in different tissues. Could those spots be nuclei? The two men left their coffee half-finished and rushed off to Schwann's laboratory to look at his slides.

Schleiden published first. *All plants are formed of cells.* Schwann was close behind. *All animals are formed of cells.* This was the cell theory in a nutshell, and at last biology had found its fundamental law. Material objects are built of atoms; plants and animals are built of cells. It seemed likely, and soon it would be proved, that each of those tiny cells was as complex and as busy as a crammed, whirring factory. The key to life was not a vital force, which had been pursued for millennia and never seen. *Cells* were the hallmark of life.

One last, crucial insight came two decades later. *"All cells come from cells."* So declared a German physician named Rudolf Virchow, in 1858, and now all the building blocks of the theory of building blocks stood in place. In particular, the nature of sperm and egg finally came into focus. If spermatozoa were in fact sperm cells and the egg was a cell, too, then these two mysterious structures were finally on a par. The Hundred Years War over which one was truly important—and this war, like the real one, had actually lasted *more* than a hundred years— could end in a peace treaty that gave equal weight to both sides.

Better still, if Virchow was right that "all developed tissues can be traced back to a cell," then conception finally made sense. It was not, as the ovists had insisted, that the embryo was hidden inside the egg, and semen merely spurred the egg to action. Nor was it, as the spermists claimed, that the embryo was hidden inside the spermatozoa, and the egg simply provided food to nourish it. Instead, the true picture must be that sperm cell and egg cell somehow merged into a single, new cell, which in turn divided and grew, as did its descendants and countless more generations of descendants, until a tiny embryo became a complex, multicellular, living being.

T HESE THREE YOUNG GERMANS WERE ONLY THE MOST PROMI-nent among a host of colleagues and rivals, but the cell theory they helped shape in the mid-1800s is still taught in the opening days of every biology class today. Insight might have come far sooner, but happenstance and bad fortune conspired to trip up all the early investigators. The main problem was that microscopes weren't up to the task. Robert Hooke had coined the term "cell" as far back as 1665, when he had examined a slice of cork and seen "a great many little Boxes" that reminded him of the cells in a beehive. Gazing at those empty boxes, Hooke saw why cork was so light in weight, but he was looking at a slice of dead wood, and he did not guess that a living cell is actually a frenetic workplace and not merely a geometric shape.

All through the 1700s, microscopes remained crude and difficult to use (and, as we have seen, unpopular). On top of that, scientists had focused their gaze in the wrong direction. Plant cells are easier to see than animal cells because plants have thick, sturdy cell walls, while animals have flimsier membranes; roughly speaking, the difference is that between a cardboard box and a plastic bag. But in the 1700s, almost no one was staring at plants under a microscope. Botanists spent their time on classification, not microscopy, and most scientists preferred working on animals in any case. Finally, in the early decades of the 1800s, microscopes improved, and plant biology came back into vogue.

Even so, the acceptance of cell theory was decades in coming. In hindsight, Virchow's declaration that *Omnis cellula e cellula* ("all cells come from cells") rings with the authority of *E pluribus unum*. At the time, it was one controversial view among many. But at the least, it would not be ignored. Rudolf Virchow would see to that.

Virchow, who became the most prominent of the cell theory pioneers, was brilliant, opinionated, often wrong-headed, and always in the middle of a dozen far-flung activities at once. He fought in the streets of Berlin in the Revolution of 1848; he was the first person to realize that cancer arose when a healthy cell turned rogue; he explored the ruins of Troy with Heinrich Schliemann; he vigorously opposed the theory of evolution (while claiming, in a half-hearted stab at good manners, that he did not mean to imply that Darwin and his followers were "stark fools and idiots"); he debunked the thrilling discovery of a Neanderthal skull and insisted that in fact it belonged to a modern man whose head someone had bashed in.

Controversy was meat and drink to Virchow. (Even his medical views reflected his combative nature; cancer was, for instance, "civil war between cells.") A crusader and a political liberal, he led campaigns for clean water and safe food. For thirteen years he served in the German Reichstag. At one point Virchow so enraged Otto von Bismarck (by insisting that the military budget was too high) that the Iron Chancellor challenged him to a duel. Virchow, because he was the man challenged, had the right to pick the weapons. He chose sausages, one raw and infested with the organisms that cause trichinosis, for Bismarck to eat, and one safe and cooked, for himself. Bismarck withdrew his challenge.

The fight over cell theory was so prolonged partly because it made bold, sweeping claims and partly because it shone a spotlight on all the still unsettled questions about the existence of a "life force." Theodor Schwann lined up proudly in the pro-mechanist, anti-vital force camp. Life was chemistry, he proclaimed, not auras and pixie dust. Living organisms followed "blind laws . . . like those of inorganic nature, which are established by the very existence of matter." The way forward was clear. *Don't bother with "forces." Find out how cells operate.*

Find out, in particular, how sperm and egg cells operate. That goal, so elusive for so long, now seemed tantalizingly within reach. The first to grab the prize, in 1875, was a vain and grumpy German scientist working at a laboratory in Naples, Italy. Oscar Hertwig would not have figured on anyone's list of likely sexual pioneers. Hertwig was frosty and forbidding, a small man with a neat beard, a bald head, a brilliant mind, and a vast disdain for virtually everyone except his brother, Richard, his coauthor on a host of scientific papers.

Many of those papers had to do with sea urchins, which themselves hardly leap to mind as creatures whose sex lives would repay close attention. But the pairing of the short-tempered scientist and the spiny sea creatures that thrived in the Bay of Naples would make biological history.

Hertwig had come to Naples to work at a newly established marine research station. Early on, the scientists there had not yet decided which creatures would make the best research subjects. Sea urchins rose to fame by accident, the marine counterparts of those aspiring actors who one day happened to serve drinks to a Hollywood mogul.

Local fishermen, who liked to gulp down sea urchin eggs, had provided the clue. The eggs were delicious (as admirers of *uni* can attest), and they were abundant (even from a single urchin). More important, they were see-through. You could put them under a microscope and see inside, almost as if you had found a peephole onto a construction site. Easy-to-harvest, transparent eggs were a scientific godsend, all the more so since you could dine on any extra experimental subjects you might have gathered.

On a spring day in 1875, Hertwig stared into his microscope at a sea urchin egg. For sea urchins, as for frogs, fertilization is external. Within the clear egg, the nucleus was distinct and unmistakable. Hertwig poked a drop of sea urchin semen near the egg. A tiny sperm cell pushed against the egg's outer surface. Moments later the nucleus of the sperm cell came into view, inside the egg, like a message thrust inside a bottle. Somehow the nucleus of the sperm cell traveled through the giant egg, making its way toward the nucleus of the egg.

Suddenly the two nuclei were in contact, and then—before Hertwig's eyes—the two nuclei fused into one. No one in history had ever seen the process of fertilization play out. Until Hertwig. The emergence of a single nucleus where there had been two, he wrote, in perhaps the only lapse into poetry in his long career, "arises to completion like a sun within the egg."

HERTWIG NOTED, TOO, THAT A SINGLE SPERM FERTILIZES A single egg. The next order of business was sorting out what goes on when a fertilized egg divides. Here, again, Hertwig helped lead the way. He and his fellow embryologists spent hour upon hour watching, fascinated, as one cell became two, those two became four, and so on. All this was terribly complicated, even in principle, since each tiny cell included a vast array of motors and assembly lines working at top speed, a fully automated Boeing airplane factory churning away in a space the size of the dot over the letter *i* in the word *impossible*.

How could it be that each cell contained its own complicated factory? A fertilized egg started as a single cell. When that cell divided in two, did each of the two "daughter" cells receive half the original machinery? Or did all the machines and tools somehow replicate themselves and then move to their new homes?

Both alternatives seemed inconceivable. *Machines that made copies of themselves? Machines that could be split in two but that kept working nonetheless?* Worse yet, this was not a one-time miracle but an endless series of miracles. That first fertilized cell gave rise to billions upon billions of new cells, each of them unimaginably intricate and complex. And the cell choreography proceeded on two levels simultaneously. Cells interacted with one another in intricate ways, and at the same time, countless components inside each cell carried on a high-speed dance.

A zoologist named Hans Driesch, at Naples, devised an ingenious experiment to probe these changes. Driesch was snobbish and made a hobby of cultivating enemies, but no one disputed his scientific talent. His idea was to wait for a sea urchin embryo to divide into two cells and then tease those two cells apart, ever so gently. What would happen? Driesch kept watch. He might have guessed that both cells would die. Both lived. He might have guessed that they would live, and one cell would develop into half of an adult sea urchin and the other cell into the other half. They didn't.

Instead, both cells grew into complete, intact, sea urchin adults, sound in mind and body. Driesch tried a variation, this time letting the embryo reach the four-cell stage and then tweaking it apart into four. Again, each of the four cells grew into a complete, functioning adult.

Observing was one thing, understanding another. Driesch's findings set off a battle among biologists. (Driesch, always happy to outrage his peers, insisted that he had demonstrated the existence of the much-reviled "life force." Life, he declared, contained mysteries that mere chemistry would never explain.) The story grew ever more complex. After several more cell divisions, individual cells lost the ability to give rise to entire organisms. Instead, cells specialized. In mammals, for instance, they took on the roles of bone and brain and heart and hair.

Unraveling the mysteries of cell growth and cell division would take decades; the saga played out over the first half of the twentieth century. The path of discovery would wend through genes and chromosomes and DNA, the landmarks in the history of modern biology. How those chromosomes governed identity and how they divvied themselves up, inside sperm and egg, and then how they spliced themselves together inside an embryo—all those discoveries lay ahead. The resolution of the mystery was a revelation: it wasn't just the *machinery* in the cell that was passed along when a cell divided but *instructions* for constructing a whole array of new machines.

No one had dreamed of such a thing in 1677, when Antony van Leeuwenhoek leapt from his marriage bed, or in 1827 when Karl von Baer stared dumbfounded at an egg from a dog's ovary, or in 1875 when Oscar Hertwig saw sperm and egg fuse into one. Those pioneering scientist-detectives had not solved the mystery themselves, but they had, at least, found the crucial clues that would enable their successors to close the case.

WE WILL NEVER KNOW WHAT ANCIENT ANCESTOR OF OURS FIRST asked where babies come from. Was it a new mother who had sweated and screamed as she pushed a baby into this harsh world? Or perhaps a sage elder staring into a fire and pondering unknowable riddles? Or an alert six-year-old watching her helpless, brand-new sister?

All of them, and countless of our forebears, would have looked around them at babies crying and nursing and gurgling and howling, and birds flying, and bugs scurrying, and wondered how those marvelous, maddening creatures came to be. Then a spark from the fire or a clap of thunder would have distracted them, and they would have stopped wondering and gone on with their lives.

So do we all. Familiarity drains the surprise from events that should make our jaws drop in disbelief. We think there is something magical about getting a rabbit out of a hat, the writer John Stewart Collis once observed. Not so. The real magic is getting a rabbit out of a rabbit.

ACKNOWLEDGMENTS

WE STROLL ALONG A PATH HACKED OUT WITH ENORMOUS LABOR BY OUR intellectual ancestors, but it is easy to neglect their achievements. The distinguishing feature of science, the anthropologist Max Gluckman remarked, is that "the fool of this generation can go beyond the point reached by the genius of the last generation." It is vital to remember that they *were* geniuses, despite their eccentric decision to have lived so long before our own enlightened age. We are all prone to a kind of present-day provincialism; I have tried hard to guard against it. To venture into the past with these bewildered, determined explorers is to marvel at the expanse of the deep forest and to gain respect for those who managed, eventually, to find paths through it. I hope I have conveyed both their genius and their befuddlement.

I relied for guidance on the scientific explorers themselves and on accounts by a host of present-day scientists and historians. I'm especially grateful to the research staffs at the New York Public Library and the American Museum of Natural History. I owe special thanks, also, to Douglas Anderson, a historian I have never met. Anderson has put together a marvelous website called Lens on Leeuwenhoek. For anyone with an interest in the history of science, the site is indispensable.

I'm relieved, too, no longer to have to face the sneers of the postman, as he delivers yet another manual on seventeenth-century sexual practices.

Alison MacKeen, Ben Platt, and Leah Stecher provided wise advice on topics both tiny and sweeping. Beth Wright copyedited with a deft hand and a sharp eye. Flip Brophy, my agent and my friend, has more verve and pizzazz than a dozen lesser souls; she embraced and guided this project from Day 1. Over countless cups of coffee throughout the course of a twenty-year breakfast seminar, my friend Al Singer provided a model of the scientific mind at work. For his challenges, insights, and, above all, encouragement, I owe him an enormous debt.

My two sons, one a novelist and the other an editor, weighed in on countless decisions. No writer could have better allies.

Lynn deserves more thanks than I can put in words.

ILLUSTRATION CREDITS

NOTES

Sources for quotations and for assertions that might prove elusive can be found below. To keep these notes in bounds, I have not documented facts that can be readily checked in standard sources. Publication information is provided only for books and articles not listed in the bibliography.

Wherever possible, I have cited works in their easiest-to-find form. For anything and everything to do with Antony van Leeuwenhock, in particular, the marvelously organized and thorough website Lens on Leeuwenhoek (http://lensonleeuwenhoek.net) is invaluable.

PROLOGUE

1 **notoriously "hot-headed"**: Aubrey, *Brief Lives*, 145.
2 **he was crack-brained"**: Ibid.
2 **the idioms are fossils**: Wright, *Harvey*, 59.
2 **"Thy *Observing* Eye first found**: Ibid., 224.
3 **for Recreation and Health sake**: Keynes, *Harvey*, 343.
3 **"much delighted in this kind**: Ibid.
3 **"no seed at all**: Ibid., 345.
4 **a "control" group**: James Lind, who showed in 1747 that lemons prevent scurvy, is often cited as the first to carry out a controlled experiment. Others credit Francesco Redi, for experiments in 1666 showing that flies do not spontaneously generate within meat scraps, or Lazzaro Spallanzani, who showed in 1768 that microorganisms do not spontaneously generate inside sealed containers.
4 **"They peremptorily did affirm**: Keynes, *Harvey*, 345.

CHAPTER ONE: ONWARD TO GLORY

7 **Isaac Newton did not know**: Home, "Force, Electricity, and the Powers of Living Matter," 112.
7 **they were parasites**: Gasking, *Investigations*, 51.
12 **the scientists' "cold philosophy"**: "Do not all charms fly / At the mere touch of cold philosophy?" Keats had written, in *Lamia*, in 1820.
12 **"Perhaps it is not surprising"** [FN]: Montesquieu, *Persian Letters*, no. 59. Online at http://tinyurl.com/hw2d8bo.

12 **God "took delight to hide his works":** William Thomas Smedley, *The Mystery of Francis Bacon* (London: Robert Banks, 1912), 104.

14 **"262 groundless hypotheses":** Jocelyn Holland, *German Romanticism and Science: The Procreative Poetics of Goethe, Novalis, and Ritter* (New York: Routledge, 2009), 5.

Chapter Two: Hidden in Deep Night

15 **prom dresses, jealousy, chamber music?":** Robert Krulwich pondered "the Hard Problem" and quoted Diane Ackerman in an NPR essay called "Building Me: A Puzzlement." The Ackerman passage is from her *The Moon by Whale Light* (New York: Random House, 1991), 131. Krulwich's essay is online at http://tinyurl.com/hjys4br.

16 **Aeschylus had spelled it out:** The passage (italics in original) is from *The Eumenides*, lines 666–671. This translation is from *The Oresteia: Agamemnon, The Libation Bearers, The Eumenides* by Aeschylus, translated by Robert Fagles (New York: Penguin Classics, 1984).

17 **"The woman hath a womb:** Helkiah Crooke, *Microcosmographia*, quoted in Aughterson, ed., *Renaissance Woman*, 55.

17 **Nature, lamented William Harvey:** Keynes, *Harvey*, 337.

17 **An egg outweighs the sperm:** Diamond, *Why Is Sex Fun?* 21.

18 **But not *altogether* impossible** [FN]: Wright, *Harvey*, 125–126.

18 **"You might be stopped by your disgust":** Quoted in Jones, *The Lost Battles*, 4.

19 **"Who would have solicited:** De Graaf, *Reproductive Organs*, 49.

19 **Scientists today** [FN]: Robert Trivers, *The Folly of Fools: The Logic of Deceit and Self-Deception in Human Life* (New York: Basic Books, 2011), 99.

20 **oddly packaged counterweights:** Aristotle, *On the Generation of Animals*, Book I, section 4. Online at http://tinyurl.com/jsjqwmp.

20 **menstrual blood soured wine:** Bainbridge, *Making Babies*, 77.

21 **two heads and four arms:** Moore cites these questions in *Science*, 236.

21 **females as "mutilated males.":** Aristotle, *On the Generation of Animals*, Book II, section 3. Online at http://tinyurl.com/jsjqwmp.

21 **William Harvey's confidence:** Wright, *Harvey*, 173.

22 **Scholars debated why God** [FN]: Marjorie Hope Nicolson, *Mountain Gloom and Mountain Glory: The Development of the Aesthetics of the Infinite* (Seattle: University of Washington Press, 1997), 59–62 and 88–91.

22 **"What was God trying:** Pinto-Correia, *The Ovary of Eve*, 14.

22 **"nothing was certain:** Roger, *Life Sciences*, 38.

Chapter Three: Swallowing Stones and Drinking Dew

23 **In the ancient world** [FN]: Riddle, *Contraception and Abortion*, 5.

24 **sunlight, moonlight, rainbows:** The examples in this paragraph and the next come from Stith Thompson, *Motif-Index of Folk-Literature: A Classification of Narrative Elements in Folktales, Ballads, Myths, Fables, Mediaeval Romances,*

Exempla, Fabliaux, Jest-Books, and Local Legends (Bloomington: University of Indiana Press, 1955), 391–395. Online at http://tinyurl.com/h4ou9ad.

24 **(The advice came from Pliny:** Pliny the Elder, *Natural History*, vol. 29, chap. 27. Online at http://tinyurl.com/hf8vnjq.

24 **"feces of crocodile:** Riddle, *Contraception and Abortion*, 66. Riddle also discusses elephant-dung suppositories.

24 **"Take dandruff from the scalp:** Manniche, *Sexual Life in Ancient Egypt*, 104.

26 **"fond of nocturnal Embraces,":** *The Works of Aristotle, In Four Parts* (London, 1777), 26. Online at Google books.

26 **The *Masterpiece* tended to the vague:** The examples in this sentence are from *Works of Aristotle*, 27.

26 **Woodcuts showed such "monsters":** *Works of Aristotle*, 68.

27 **"Subtle Lechers! Knowing that:** The book-length poem was called "Kick Him Jenny" and was published in 1735.

27 **"Dwarfs, Cripples, Hunch-backed:** Nicolas Venette, *Conjugal Love Reveal'd*, quoted in McLaren, *Reproductive Rituals*, 45.

28 **"When the woman is on top:** Both quotations in this sentence come from Flandrin, *Sex in the Western World*, 120.

28 **the Trobrianders remained "entirely ignorant":** Malinowski, "Baloma," 220. Malinowski's discussion of the Trobrianders' theory of conception is in Chap. 7, with occasional remarks in Chaps. 1, 2, 3, 6. Online at http://tinyurl.com/z2vxxuq.

30 **anthropologists have fought:** See, for instance, *Arguments About Aborigines* by Lester Richard Hiatt.

31 **"gave birth to their spouses:** Inhorn, *Quest for Conception*, 54.

31 **"I created on my own:** Ibid., 55.

31 **A second papyrus depicts:** Ibid. The full papyrus in the British Museum can be viewed online at http://tinyurl.com/zhun748.

31 **"the !Kung believe:** Lorna Marshall, *Nyae Nyae !Kung Beliefs and Rites* (Cambridge, MA: Peabody Museum Press, 1999), 117.

32 **"as rennet acts upon milk":** Aristotle, *On Generation*, Book II, part 4. Online at http://tinyurl.com/jsjqwmp.

32 **the modern-day Basques:** Marten Stol, *Birth in Babylonia and the Bible: Its Mediterranean Setting* (Groningen, Netherlands: Styx, 2000), 15.

32 **the Bantu of southern Africa:** Albert I. Baumgarten, ed., *Self, Soul, and Body in Religious Experience* (Leiden, Netherlands: Brill, 1998), 12.

32 **"The father provides the white seed:** Kottek, "Embryology." Kottek cites similar ideas in ancient Hindu works.

32 **The historian (and embryologist) Joseph Needham:** Needham, *Embryology*, 26.

32 **"putting captured males to death:** Ibid.

33 **"If you plant wheat:** Carol Delaney, "The Meaning of Paternity and the Virgin Birth Debate," *Man*, New Series 21, no. 3 (Sept. 1986): 497.

33 **"Here in Egypt:** Inhorn, *Quest for Conception,* 70.

33 **"Many of my New Guinea friends:** Diamond, *Why Is Sex Fun?,* 65.

33 **somewhat like a snowball:** Beckerman and Valentine, eds., *Partible Paternity,* 10.

33 **So demanding is this task:** Ibid.

33 **"a good mother will make a point:** Yuval Harari, *Sapiens: A Brief History of Humankind* (New York: Harper, 2015), 39.

34 **The Talmud even spelled out:** Leo Auerbach, *The Babylonian Talmud in Selection* (New York: Philosophical Library, 1944), 162. Online at http://tinyurl .com/j7e6jbv.

34 **"Shameful kissing and touching":** Flandrin, *Sex in the Western World,* 123, and Laqueur, *Solitary Sex,* 138 and 153–154.

34 **"Adulterous is also the man:** Flandrin, *Sex in the Western World,* 4.

34 **The view was a long time** [FN]: Alice Morse Earle, *The Sabbath in Puritan New England* (New York: Scribner's, 1896), 247. Online at http://tinyurl.com/ hevxkvq.

35 **Questions that seem ludicrous:** Stephen Jay Gould looked deeply into "Adam's Navel," and Jacques Roger discussed the riddle of carnivores and herbivores in Eden in *Life Sciences,* 168. For lust in Eden, see Saint Augustine below.

35 **"ready servant of the will.":** Saint Augustine, *The City of God,* Chap. 23. Online at http://tinyurl.com/gpsx7la.

35 **"rather like a drawbridge":** Jacobs, *Original Sin,* 61.

35 **"tranquility of mind":** Saint Augustine quoted in Nightingale, *Once Out of Nature,* 30.

35 **"The female parts:** Ibid., 46.

36 **set the clock forward:** Ibid., 30.

36 **devoured by sharks:** See the brilliant essay called "Continuity, Survival and Resurrection," Chapter 7 in Bynum, *Fragmentation and Redemption.* See also John Carey, *John Donne: Life, Mind and Art* (New York: Oxford University Press, 1981), 219–226.

36 **"the eaten flesh will be restored:** Saint Augustine, *The City of God,* Book 22, Chap. 20. Online at http://tinyurl.com/zhsabbc.

36 **"All our activity:** Nightingale, *Once Out of Nature,* 51.

36 **To each his own** [FN]: Mark Twain, *Notebook,* 397. Online at http://tinyurl .com/jk24f96.

36 **"In order that the happiness:** Thomas Aquinas, *Summa Theologica,* vol. 5 (Part III, Second Section and Supplement), 2960. The quoted passage comes from a section entitled, "Whether the Blessed in Heaven Will See the Sufferings of the Damned?"

36 **"What bliss will fill:** Donald Bloesch, *The Last Things: Resurrection, Judgment, Glory* (Downers Grove, IL: InterVarsity, 2004), 223.

CHAPTER FOUR: UNMOORED IN TIME

37 **The very word "autopsy":** Bainbridge, *Making Babies*, 56.

38 **From roughly 1300:** Park, "The Criminal and the Saintly Body," 4–6.

38 **Cleopatra supposedly ignored:** Needham, *Embryology*, 47.

38 **Nero, seldom outdone:** Park, "Dissecting the Female Body," 29–38. Park notes that the story first appeared more than a thousand years after Nero's death.

39 **the "great disgust" we feel:** Andrea Carlino, *Books of the Body: Anatomical Ritual and Renaissance Learning* (Chicago: University of Chicago Press, 1999), 156.

39 **He missed another chance** [FN]: Susan Mattern, *The Prince of Medicine: Galen in the Roman Empire* (New York: Oxford University Press, 2013), 222.

40 **"windows into the body":** Rob Dunn, *The Man Who Touched His Own Heart: True Tales of Science, Surgery, and Mystery* (New York: Little, Brown, 2015), 29.

40 **"It is far more excellent:** Saint Augustine, *On the Soul and Its Origin*, Chap. 14. Online at http://tinyurl.com/jrvwxq4.

40 **"cruel zeal for science":** Carlino, *Books of the Body*, 165.

40 **"the lust of the eyes":** William Eamon, *Science and the Secrets of Nature: Books of Secrets in Medieval and Early Modern Culture* (Princeton, NJ: Princeton University Press, 1996), 60.

41 **"If the wisest men:** Westfall, *Science and Religion*, 22.

41 **After the Fall:** Thomas, *Man and the Natural World*, 17.

41 **"Again will Achilles go:** Wootton, *The Invention of Science*, 75.

41 **"To 'dis-cover' was to pull away:** McMahon, *Divine Fury*, 4.

42 **"Virtually every drawing:** Clayton and Philo, *Leonardo*, 9.

42 **almost precisely the same time:** Leonardo's cutaway drawing of a man and woman having sex "dates from about 1492–4," according to Keele in *Anatomical Drawings*, 69. Clayton and Philo date the drawing at "around 1490," in *Leonardo*, 10.

43 **"an hundred things:** Peter Harrison, *The Bible, Protestantism, and the Rise of Science* (New York: Cambridge University Press, 2001), 82. Harrison cited a passage from Boyle's *The Christian Virtuoso* in *Works*, vol. 5, 520.

43 **a kind of P.S.:** Kemp, ed., *Leonardo on Painting*, 251.

43 **A man of "supernatural" beauty:** Giorgio Vasari, *The Lives of the Artists* (New York: Oxford University Press, 1998), 284.

43 **pinks and purples:** Jones, *Lost Battles*, 13, 149, 157.

43 **Leonardo's mirror-writing:** Clayton and Philo, *Leonardo*, 9.

44 **As Leonardo drew things:** Keele, *Anatomical Drawings*, 69.

45 **"sperm is a drop of brain.":** The ancient writer was Diogenes Laertius (who was not the more famous Diogenes), quoted in Pieter Willem Van der Horst, *Hellenism, Judaism, Christianity: Essays on Their Interaction* (Kampen, Netherlands: Kok Pharos, 1998), 221.

45 **"soft and feeble":** Keele, *Elements*, 350.

45 **"beans and Pease:** Jane Sharp, *The Midwives Book or The Whole Art of Midwifry Discovered* (New York: Oxford University Press, 1999), 30.

45 **"If an adversary says:** Keele, *Elements*, 350.

45 **a great wind to enlarge:** Clayton and Philo, *Leonardo*, 157.

46 **"a tan-colored small cap":** Jones, *Lost Battles*, 16.

46 **"Many die thus:** Keele, *Elements*, 350.

47 **"The act of coitus:** Keele and Roberts, *Anatomical Drawings*, 69.

47 **The comment turns up** [FN]**:** Clayton and Philo, *Leonardo*, 99.

47 **"It remains obstinate:** Keele, *Elements*, 350.

48 **"This old man:** Clayton and Philo, *Leonardo*, 18.

48 **"more than thirty":** Ibid., 30.

48 **"inhuman and disgusting:** Ibid., 21.

48 **"Get hold of a skull":** "Previously Unexhibited Page from Leonardo's Notebooks Includes Artist's 'To Do' List," Royal Collection Trust, 2012. Online at http://tinyurl.com/j6mfvsz.

49 **"These I intend to describe:** Keele, *Elements*, 36.

49 **"It is noteworthy":** Ibid.

50 **"You who say it would be better:** Jones, *Lost Battles*, 3.

51 **much that is mysterious":** Suh, ed., *Leonardo's Notebooks*, 181.

CHAPTER FIVE: "DOUBLE, DOUBLE TOIL AND TROUBLE"

52 **some further terror:** Stuart Banner, *The Death Penalty: An American History* (Cambridge, MA: Harvard University Press, 2009), 77.

53 **"Evil men, who did harm:** C. Jill O'Bryan, *Carnal Art: Orlan's Refacing* (Minneapolis: University of Minnesota Press, 2005), 65.

53 **"secrets of nature revealed:** Gross, "Rembrandt's 'The Anatomy Lesson.'"

54 **unmarried and abandoned:** Julie V. Hansen, "Resurrecting Death: Anatomical Art in the Cabinet of Dr. Frederick Ruysch," *Art Bulletin* 78, no. 4 (Dec., 1996): 671.

54 **"Mr. Doctor be made not:** Power, *Harvey*, 58.

54 **stealing hearts or kidneys:** Heckscher, *Rembrandt's Anatomy*, 27.

54 **"in order that everyone may come":** Wright, *Harvey*, 61.

55 **"First you must obtain:** Vesalius, *Human Body*, Book I, 371.

55 **"in the hope of seeing:** Ibid., Book I, 382.

57 **"troublesome, dirty, and difficult.":** Ibid., Book I, 370.

57 **Macbeth's witches:** All quotes in this paragraph are from Vesalius, *Human Body*, Book I, 374.

58 **Samuel Pepys wrote:** Pepys's diary is online at http://tinyurl.com/kcbu4qt.

58 **"Jews or other infidels.":** Wright, *Harvey*, 71.

59 **So-called resurrection men:** Stott, *The Poet and the Vampyre*, 33.

60 **a murder like this:** Rosner, *The Anatomy Murders*, 1. The full passage can be found in *West Port Murders* by Thomas Ireland (Edinburgh, 1829), 1.

60 **Dr. Knox claimed:** Bates, *Robert Knox*, 69.

62 **"storm-tossed upon a mighty sea:** Vesalius, *Human Body*, Book V, 145.

62 **Genius though he was:** Fritjof Capra, *Learning from Leonardo: Decoding the Notebooks of a Genius* (San Francisco: Berrett-Koehler, 2013), 294.

62 **"The unsuspecting student plunges:** Miller, *The Body in Question*, 177.

Chapter Six: Door A or Door B?

64 **"A Man was first a Boy":** Aughterson, ed., *Renaissance Woman*, 406.

65 **"But he had not read very long":** Aubrey, *Brief Lives*, 131.

65 **"His tongs were ready:** Keynes, *Harvey*, 214.

66 **"which is itself loathsome":** Ibid., 96.

67 **"The examination of the bodies:** Power, *Harvey*, 148.

68 **Robert Boyle** [FN]: Fudge, ed., *Renaissance Beasts*, 199. The passage in Boyle comes from his *Works*, vol. 2, 7.

68 **"we could scarce see:** Power, *Harvey*, 85.

68 **"a spleen hanging like a letter V.":** Keynes, *Harvey*, 132.

68 **according to one biographer:** Wright, *Harvey*, 98.

68 **special note of the "huge" colon:** Ibid.

68 **As well as demonstrably** [FN]: Richardson, *Death, Dissection*, 31.

70 **the historian Thomas Laqueur:** Laqueur, *Making Sex*, 79.

70 **women with their insides out:** Fletcher, *Gender, Sex, and Subordination*, 37.

71 **"Nothing is accidental:** Leroi, *The Lagoon*, 10.

72 **"When also in coition:** Nathaniel Highmore, *The History of Generation* (London: 1651), 85. Online at http://tinyurl.com/zd5qf3r.

72 **"Men should take their time:** Jacquart and Thomasset, *Sexuality and Medicine*, 130–131.

73 **one voluptuous sensation.":** Harvey, *On Generation* in *The Works of William Harvey* (London, 1857), 294. Online at http://tinyurl.com/zq2wb4e.

73 **"I, for my part, greatly wonder:** Merchant, *The Death of Nature*, 159.

73 **"pre-eminently the seat of woman's delight":** Laqueur, *Making Sex*, 64.

74 **the names of their male counterparts:** Fletcher, *Gender, Sex, and Subordination*, 35.

74 **Galen made a rueful comparison:** Laqueur, *Making Sex*, 28.

74 **By this reasoning** [FN]: Pantel, ed., *Women in the West*, 66.

74 **"Their secret internal organs:** Charles Rosenberg and Carroll Smith-Rosenberg, "The Female Animal: Medical and Biological Views of Women," in Charles Rosenberg, *No Other Gods: On Science and American Social Thought* (Baltimore: Johns Hopkins University Press, 1997), 57.

75 **"owing to youth:** Pantel, ed., *History of Women*, 75.

76 **"It rose naturally toward the heavens:** Wiesner, *Women and Gender*, 32.

76 **In the Middle Ages** [FN]: Boyce, *Born Bad*, 37.

76 **"In a weaker organism,":** Pantel, ed., *History of Women*, 66.

77 **"woman is at one and the same time:** Roger, *Life Sciences*, 46.

CHAPTER SEVEN: MISSING: ONE UNIVERSE (REWARD TO FINDER)

82 **"I myself have seen a hen ostrich:** Birkhead, *Wisdom of Birds*, 312.

83 **The eggs of all mammals:** Carl G. Hartman, "How Large Is the Mammalian Egg?" *Quarterly Review of Biology* 4 (1929): 373–388.

83 **"In a black drake:** Birkhead, *Wisdom of Birds*, 312.

83 **a rooster burned at the stake:** Ibid., 274.

84 **"an almost complete egg:** Power, *Harvey*, 29.

84 **"as by Contagion":** Wilson, *Invisible World*, 110.

84 **"epidemic, contagious, and pestilential:** Merchant, *The Death of Nature*, 160.

85 **"analogous to the essence:** Wilson, *Invisible World*, 107.

85 **An embryo was a "conception":** Moore, *Science as a Way of Knowing*, 484, and Bainbridge, *Making Babies*, 65. Shakespeare punned on the double meaning of "conceive" in *King Lear* (Act 1, Scene 1).

85 **When the *New York Times* reported:** *New York Times*, Dec. 29, 1960.

86 **now called capillaries:** Bainbridge, *Making Babies*, 59. The first to see capillaries was Antony van Leeuwenhoek, in 1688. (Harvey died in 1657.) See Ruestow, *Microscope*, 175.

86 **"mucous filaments like spiders' threads,":** Keynes, *Harvey*, 346–347, and Cobb, *Generation*, 28–29.

86 ***Disputations Touching the Generation:*** The book was originally published in Latin but translated almost at once into English.

86 **"till he was almost dead with cold,":** Aubrey, *Brief Lives*, 134.

87 **Those who would reduce:** Ibid.

87 **"All animals whatsoever:** Trounson and Gosden, eds., *Oocyte*, 3.

87 **Both the hen and housewife:** The poem was by a writer and physician named Martin Lluelyn.

88 **he discounted the role:** Roger, *Life Sciences*, 205.

88 **did "but replace one mystery:** Gasking, *Investigations*, 35.

88 **"There is no sensible [i.e., detectable] thing:** Trounson and Gosden, eds., *Oocyte*, 5.

88 **"but to confess myself:** Gasking, *Investigations*, 35.

CHAPTER EIGHT: SHARKS' TEETH AND COWS' EGGS

89 **"the testicle of a dormouse:** Cobb, *Generation*, 122.

89 **"you will behold a delightful:** De Graaf, *Reproductive Organs*, 25.

90 **"Whatever is procreated of the semen:** Harvey, *Works*, 171. Online at http://tinyurl.com/zq2wb4e.

91 **"the part that enters:** De Graaf, *Reproductive Organs*, 10.

91 **"The pleasure of copulation:** Ibid., 47.

91 **"a citizen of Delft:** Ibid., 13.

92 **they "could not possibly act:** Ibid., 28.

92 **"I am certainly surprised:** Ibid., 34.

92 **"Your book came out:** Cobb, *Generation*, 183.

92   **the fact that milk is white:** De Graaf, *Reproductive Organs*, 32.
92   **"twisted like a waxen nose":** Cobb, *Generation*, 120.
92   **"I pray thee, O God:** Cutler, *Seashell*, 27.
93   **if fleas have bones:** Ibid., 33.
94   **"Having seen that the testicles:** Cobb, *Generation*, 99–100.
94   **as if a warring host:** Ruestow, *Microscope*, 118.

CHAPTER NINE: THE EGG, AT LAST

96   **"made bold to strip:** De Graaf, *Reproductive Organs*, 79.
97   **"The notion of some people:** Ibid., 106.
97   **"certain females . . . with lascivious thoughts:** Ibid., 107.
97   **"In libidinous women":** Ibid., 141.
97   **"The woman's vagina is so cleverly:** Ibid., 107.
97   **a benign fluid:** Ibid., 132.
97   **The size of the nose:** Ibid., 46
97   **both "claimed the glory":** Ibid., 89.
97   **"a most unfortunate topography:** Gonzalez-Crussi, *Carrying the Heart*, 151.
98   **"set between the bladder:** De Graaf, *Reproductive Organs*, 110.
98   **a pear which has been slightly squashed,":** Ibid.
98   **"Writers of the sharpest wits:** Ibid.
98   **all of them "have ovaries full of eggs:** Ibid., 81.
98   **"Since all this can be observed:** Ibid., 82.
98   **"frivolous and stupid":** Ibid., 81.
98   **"like the peacock's tail":** Ibid., 9.
99   **"Nature had her mind on the job:** Ibid.
99   **Bias against females** [FN]**:** Merchant, *The Death of Nature*, 157, and Pinto-Correia, *The Ovary of Eve*, 41.
99   **"I am confident that thousands:** Elizabeth Potter, *Gender and Boyle's Law of Gases* (Bloomington: Indiana University Press, 2001), 6.
99   **"The common function of the female 'testicles':** De Graaf, *Reproductive Organs*, 135.
100 **(Vesalius, anatomy's founding father:** Gasking, *Investigations*, 37.
100 **"had been disrupted or expelled":** De Graaf, *Reproductive Organs*, 166.
100 **To his delight:** Ibid., 82.
101 **"come to a sinister end.":** Ibid., 167.
102 **"seminal vapor" and "irradiation":** Ibid., 149, 81.
102 **lamenting "the disaster:** Cobb, *Generation*, 179.
103 **"I am writing to tell you:** The letter is online at http://tinyurl.com/j4puk45.

CHAPTER TEN: A WORLD IN A DROP OF WATER

104 **"very many little animalcules":** Dobell, *Leeuwenhoek*, 110.
105 **"The motion of most of the animalcules:** Ibid., 111.
105 **"flies which looked as big as a lamb":** James Newman, ed., "Commentary

on Galileo Galilei," *The World of Mathematics* (New York: Simon and Schuster, 1956), 2:732n.

105 **"like an Iron bar:** Hooke, *Micrographia.* Online (in a beautifully produced version) at http://tinyurl.com/zdqxtwl.

105 **"little animals more than a thousand times less:** Dobell, *Leeuwenhoek*, 121.

106 **"saliva, chyle, sweat, etc.":** Wilson, *Invisible World*, 131.

106 **"I felt averse:** Ibid.

106 **He nagged her for blood:** Ruestow, *Microscope*, 156.

106 **He pestered shopkeepers:** Ibid.

106 **"so small that I judged:** Dobell, *Leeuwenhoek*, 133.

107 **"We had such stories written:** Wilson, *Invisible World*, 237.

107 **Magnifying glasses had been known:** Vincent Ilardi, *Renaissance Vision from Spectacles to Telescopes* (Philadelphia: American Philosophical Society, 2007), 4, 42.

108 **In hindsight the path:** Steven Johnson, *How We Got to Now: Six Innovations That Made the Modern World* (New York: Riverhead, 2014), 22–24.

108 **Leeuwenhoek's first encounter:** Snyder, *Eye of the Beholder*, 55.

109 **Eight prominent citizens:** Lens on Leeuwenhoek. Online at http://tinyurl .com/hhxmvkw.

109 **"No discovery made":** For this quotation and those in the following two paragraphs, see http://tinyurl.com/zdvpznb.

110 *zaad-ballen*, **"seed-balls.":** Lens on Leeuwenhoek. Online at http://tinyurl .com/hcvofbt.

111 **"as thick upon his Carcass":** Letter written on June 21, 1701. Online at http://tinyurl.com/jra6gtx.

111 **He nearly blinded himself:** Dobell, *Leeuwenhoek*, 6.

111 **"did not smell at all good:** Letter written on July 22, 1704. Online at http:// tinyurl.com/jgnv4dy.

111 **"an hungry Lowse" perched:** Letter written on August 15, 1673. Online at http://tinyurl.com/h6f3fa9.

111 **"carried in her Bosom:** Letter written on July 11, 1687. Online at http:// tinyurl.com/hrn8oua.

111 **"a little white matter":** Letter written on Sept. 17, 1683. Online at http:// tinyurl.com/gon2p4d.

111 **"Some of them were so disgusted:** Ibid.

112 **"there are not living in our United Netherlands:** Ibid.

112 **"divisions [on a brass ruler]:** Letter written on Nov. 12, 1680. Online at http://tinyurl.com/jq9akz5.

112 **"I also with much trouble:** Letter written on Jan. 22, 1683. Online at http:// tinyurl.com/gvjt4jl.

112 **he stared entranced:** Ruestow, *Microscope*, 176.

112 **less and less enticing:** Dobell, *Leeuwenhoek*, 222, and Snyder, *Eye of the Beholder*, 289.

112 **"animalcules a-moving very prettily":** Dobell, *Leeuwenhoek*, 224.

112 **"partly with his Nails:** Letter written on July 7, 1722. Online at http://tinyurl .com/hgbw8tq.

CHAPTER ELEVEN: "ANIMALS OF THE SEMEN"

114 **"immediately after ejaculation:** Letter written in November 1677. Online at http://tinyurl.com/zp4eycy.

116 **"They cause no inconvenience:** "Clinical Lecture by Dr. Elliotson, Delivered at St. Thomas's Hospital, February 1, on Intestinal Worms," *Lancet*, March 3, 1830.

116 **"animals of the semen.":** Farley, *Gametes and Spores*, 43.

116 **animated swizzle sticks:** Ibid., 46, 56.

118 **"all manner of great and small:** Letter written in Nov. 1677. Online at http:// tinyurl.com/zp4eycy.

118 **"it is exclusively the male semen:** Letter written on March 18, 1678. Online at http://tinyurl.com/zmgz9ea.

118 **"Our Harvey and your de Graaf":** Wilson, *Invisible World*, 133.

119 **"seventy times seventy more":** Letter written on March 30, 1685. Online at http://tinyurl.com/zpxmy45.

119 **"I discovered to my great satisfaction:** Ibid.

119 **"obstinately opinionated":** Ibid.

119 **"A human being originates:** Letter written on Jan. 22, 1683. Online at http:// tinyurl.com/jalkcvo.

120 **nobody understood where those seeds:** Gasking, *Investigations*, 65.

120 **"egg was sucked from the egg-nest":** Letter written on March 30, 1685. Online at http://tinyurl.com/z2sx35j

121 **"Had there been one particle:** Ibid.

121 **"addle-pated," "fantastic":** Ibid.

121 **"I have sometimes imagined:** Letter written on July 13, 1685. Online at http://tinyurl.com/hy4e44v.

121 **"The parts lying in such an animalcule:** Letter written on Dec. 25, 1700. Online at http://tinyurl.com/hyu29lr.

CHAPTER TWELVE: DOLLS WITHIN DOLLS

128 **"The Unreasonable Effectiveness:** The essay is online at http://tinyurl.com /z8gtksk.

129 **"Were men and beast made by fortuitous:** Michael White, *Isaac Newton: The Last Sorcerer* (New York: Basic 1999), 149, quoting a notebook entry of Newton's headed "Of God."

129 **the clergyman and naturalist William Paley:** Paley was a brilliant thinker who nearly anticipated Darwin, but could not believe in a God who worked in so unlikely a way. See George Johnson, "A Creationist's Influence on Darwin," *New York Times*, May 23, 2014. Paley's essay on God as the divine watchmaker, "The

Watchmaker Argument," from his book *Natural Theology*, is online at http://tinyurl.com/jgp3e5p.

129 **Then he'd set the clockwork running:** Physicists all believed in a clockwork universe, but they fought bitterly over whether God needed to fine-tune his creation (this was Newton's view) or whether it ran smoothly and eternally on its own. See Edward Dolnick, *The Clockwork Universe: Isaac Newton, the Royal Society, and the Birth of the Modern World* (New York: Harper, 2011), 310–313.

131 **Jan Swammerdam cited** [FN]: Ruestow, "Piety," 218.

132 **"In one single spermatic worm:** Smith, *Divine Machines*, 183, quoting Nicolas Andry, *De la génération des vers dans le corps de l'homme* (*On the Generation of Worms in the Human Body*).

133 **"The majesty of string theory":** Edward Witten in an interview with the journalist John Horgan. Online at http://tinyurl.com/nsbcl46.

133 **"holy awe":** Ruestow, *Microscope*, 228.

133 **It did not reveal:** Gasking, *Investigations*, 102.

133 **We must believe / What Reason tells:** Ruestow, *Microscope*, 228, quoting Henry Baker.

134 **Malpighi had examined chicken eggs:** Roger, *Buffon*, 120.

135 **So, Nat'ralists observe, a Flea:** Swift's poem is online at http://tinyurl.com/ooj6bmz.

136 **Antony van Leeuwenhoek, far too impatient:** Ruestow, "Images and Ideas," 213–214.

136 **"God, Lord and Omniscient Maker:** Letter written on Aug. 24, 1688. Online at http://tinyurl.com/zg47hdw.

136 **almost unimaginable that "in an Animalcule:** Letter written on June 23, 1699. Online at http://tinyurl.com/zooga87.

137 **"an animalcule from the male seed:** Letter written on March 30, 1685. Online at http://tinyurl.com/joxvx9o.

Chapter Thirteen: The Message in God's Fine Print

139 **"The Deity is as conspicuous:** Harrison, "Reading Vital Signs," 201, quoting Abbé Pluche, the French priest and author of *Spectacle de la Nature* (*Nature Delineated*).

139 **"Solomon in all his glory:** Noel-Antoine Pluche, *Nature Delineated* (London: 1740), 8.

140 **God is greatest in ye Least:** Ruestow, *Microscope*, 59, quoting Henry Power.

140 **"I offer you the Omnipotent:** Jorink, "Between Emblematics," 161.

141 **"I have striven:** Ruestow, *Microscope*, 119.

141 **"All God's works:** Sleigh, "Swammerdam's Frogs," 378.

141 **"crushed, and otherwise disturbed:** Cobb, *Generation*, 149.

141 **"The male Frog leaps:** Pinto-Correia, *The Ovary of Eve*, 107.

142 **"After all is finished:** The passage is from Chapter 9 of Swammerdam's *Book*

of Nature, which is called "Of the Manner in Which Snails Mutually Perform the Business of Coition."

143 **"an almost uncontrollable passion":** James Duncan, *Introduction to Entomology* (Edinburgh, 1840), 18.

143 **"What atheist who considered:** Jorink, "Between Emblematics," 236.

143 **Swammerdam delighted in:** All these examples come from Ruestow, *Microscope*, 135.

143 **"Artist of all artists":** Ruestow, *Microscope*, 137.

144 **"clearly evident,":** Jorink, "Between Emblematics," 157, quoting Johannes Godaert.

144 **"The analogy between the silkworm:** Wilson, *Invisible World*, 124.

145 **"Legs, wings, and the rest:** Pinto-Correia, *The Ovary of Eve*, 27, and Ruestow, "Piety," 227.

146 **"by a miracle which would surpass:** Gasking, *Investigations*, 37, quoting Claude Perrault.

146 **"the least reason why:** Ruestow, *Microscope*, 247.

146 **"How could hair come:** Needham, *Embryology*, 48.

147 **"see, touch, and feel God:** Ruestow, *Microscope*, 119.

147 **"heavenly reflections":** Jorink, "Between Emblematics," 153.

147 **"melancholic madness":** Ruestow, *Microscope*, 125.

CHAPTER FOURTEEN: SEA OF TROUBLES

149 **Leeuwenhoek was nearsighted:** Snyder, *Eye of the Beholder*, 281.

150 **"the patience of an angel":** Ruestow, *Microscope*, 152 fn.

150 **"on the close inspection:** Lens on Leeuwenhoek. Online at http://tinyurl .com/zsq2vbn.

151 **"These Worms are not found:** Pinto-Correia, *The Ovary of Eve*, 75.

151 **"to throw away one's books:** The rival was Denis Diderot, quoted in Matthew Stewart, *The Courtier and the Heretic: Leibniz, Spinoza, and the Fate of God in the Modern World* (New York: Norton, 2007), 12.

151 **a collection of his own sayings:** Bertrand Russell, *A History of Western Philosophy* (New York: Simon and Schuster, 1945), 582.

152 **a "Patient almost dead with Pain":** Nicholas Andry, *An Account of the Breeding of Worms in Human Bodies*, 10. Online in part at http://tinyurl.com/jn6k79n.

152 **"You shall discover in it:** Wilson, *Invisible World*, 137.

152 **"food for worms":** Shakespeare goes on at great length about worms devouring bodies in *Hamlet*, Act 4, Scene 3, after Hamlet kills Polonius, and more briefly in *Henry IV, Part One,* Act 5, scene 4, and in Sonnet 6.

153 **"Each little animal actually encloses:** Pinto-Correia, *The Ovary of Eve*, 78–79.

154 **"little men and little ladies:** Syson, *Doctor of Love*, 203.

154 **"the spermatic liquid of dogs:** Roger, *Life Sciences*, 253.

154 **an infinite number of murders:** Ibid., 251.

155 **"The womb being so large:** Letter written on Jan. 22, 1683. Online at http://tinyurl.com/gvjt4jl.

156 **James Cooke suggested that perhaps:** Moore, *Science*, 398.

157 **It was well-known** [FN]: Jean-Charles Seigneuret, *Dictionary of Literary Themes and Motifs* (Westport, CT: Greenwood, 1988), 1:670.

157 **"medicines of great efficiency":** Laqueur, *Solitary Sex*, 15.

158 **"Every seminal emission out of nature's road:** Syson, *Doctor of Love*, 207.

158 **Rousseau warned against:** *Emile*, Book 4 (1762).

158 **Kant proclaimed masturbation more sinful:** See Immanuel Kant, *The Metaphysics of Morals*, translated and edited by Mary Gregor (New York: Cambridge University Press, 2000), 179. Kant's argument was that "murdering oneself requires courage" and therefore demonstrates some "respect for the humanity in one's own person." But the masturbator, by the "complete abandonment of oneself to animal inclination," makes a person "a thing that is contrary to nature, that is, a loathsome object, and so deprives him of all respect for himself."

158 **"more like a corpse:** Tissot, *Treatise*, 19.

158 **"A pale bloody discharge:** Ibid.

159 **"the secretion which they lose:** Ibid., 45.

159 **the loss of a single ounce:** Ibid., v.

159 **the historian Jacques Roger pointed out:** Roger, "Two Scientific Discoveries," 232.

CHAPTER FIFTEEN: THE RABBIT WOMAN OF GODLIMAN

160 **Mary Toft was twenty-four years old:** This account is drawn from Todd, *Imagining Monsters*, Chapters 1–3.

161 **Soon he grew so certain:** Cody, *Birthing the Nation*, 125.

162 **"much terrified with an old lion's noise.":** Fletcher, *Gender,* 72.

162 **"the deformity which I am now exhibiting:** *The Autobiography of Joseph Carey Merrick*. Online at http://tinyurl.com/jbfppdv.

162 **twins joined at the head:** Pinto-Correia, *The Ovary of Eve*, 158.

163 **Historians have never found:** Park, *Secrets of Women*, 145.

163 **staring at paintings of their husband:** Ibid.

163 **In one particularly desperate case:** Valeria Finucci, *The Manly Masquerade: Masculinity, Paternity, and Castration in the Italian Renaissance* (Durham, NC: Duke University Press, 2003), 52. Finucci notes that the verdict was soon overturned, although some scientists continued to defend it.

165 **"I have been assured by three:** Moore, *Science*, 241.

165 **"the son of that daughter:** Aristotle, *On Generation*, Book I, part 18. Online at http://tinyurl.com/jsjqwmp. Aristotle's remark about the father and son with matching marks on their arms comes a few lines earlier in the same work.

166 **"a scoundrel who coupled:** Roger, *Life Sciences*, 19. The anecdote appeared not in an out-of-the-way account but in a work first published in 1616, then reissued in 1634, 1665, and 1668, and translated into French in 1708.

166 **Locke pondered a variety:** John Locke, *An Essay Concerning Human Understanding*, vol. 2, Book 3, Chap. 6, Section 27. Online at http://tinyurl.com/jv2ptlc.

168 **"If the woman had been:** Müller-Sievers, *Self-Generation*, 29.

168 **perhaps a tiny push:** Roger, *Life Sciences,* 306, quoting Claude Perrault.

168 **"This is so poor,":** Cobb, *Generation*, 233.

169 **Leeuwenhoek pursued the matter:** Letter written on July 16, 1683. Online at http://tinyurl.com/zth6yr9.

169 **he would have been the first:** Cobb, *Generation*, 225, and Roger, *Life Sciences*, 307.

170 **"there is no part:** Cobb, *Generation*, 131.

CHAPTER SIXTEEN: "ALL IN PIECES, ALL COHERENCE GONE"

172 **All in Pieces:** The phrase is from John Donne's poem "An Anatomy of the World."

173 **"From each portion:** Vartanian, "Trembley's Polyp," 259. For a full and lively account of Trembley's work, see Stott, *Darwin's Ghosts*, Chap. 5.

174 **"These are Truths":** Ruestow, *Microscope*, 273, quoting Henry Baker.

174 **nothing but a stomach:** Ruestow, *Microscope*, 273, paraphrasing Charles Bonnet.

174 **"If 'tis cut in two:** Stott, *Darwin's Ghosts*, 101, quoting Charles Hanbury Williams.

175 **If life could shape itself:** Dawson, "Regeneration."

176 **Bonnet kept watch "day by day:** Dawson, *Nature's Enigma,* 6.

177 **scientist named Steven Blanckaert:** Ruestow, *Microscope*, 206n.

177 **he looked closely at aphids:** Letter written on Oct. 26, 1700. Online at http://tinyurl.com/jov6erz.

177 **some mysterious "essential stuff":** Ruestow, "Images and Ideas," 221. Ruestow cites a letter Leeuwenhoek wrote to the Royal Society on July 13, 1685. Online at http://tinyurl.com/hy4e44v.

177 **to his astonishment he found:** Letter written on July 10, 1695. Online at http://tinyurl.com/j4lsztk.

178 **micro-miniature aphids:** Ruestow, *Microscope,* 206.

178 **"I was at my wits' end:** Letter written on July 10, 1695.

179 **Even the staunchest believers:** Roger, *Buffon*, 122.

179 **"He who causes himself:** Terrall, *The Man Who Flattened the Earth*, 5.

180 **he found a family in Germany:** Gregory, *Evolutionism*, 108–109. See also Gasking, *Investigations*, 80–81.

180 **"the best proven things:** Gregory, *Evolutionism*, 109.

180 **God had designed "monstrous eggs":** Roger, *Buffon*, 122.

181 **you would need a microscope:** Müller-Sievers, *Self-Generation*, 31.

182 **a metal duck, complete with copper feathers:** Riskin, "The Defecating Duck."

183 **The earliest musical notation:** Alfred Crosby, *The Measure of Reality: Quantification and Western Society, 1250–1600* (New York: Cambridge University Press, 1997), 144.

184 **Leibniz's computer was a sort of colossal:** George Dyson, *Darwin Among the Machines: The Evolution of Global Intelligence* (New York: Basic Books, 2012), 37.

CHAPTER SEVENTEEN: THE CATHEDRAL THAT BUILT ITSELF

185 **a "force that would be sufficiently wise":** Roe, *Matter, Life, and Generation*, 29.

186 **"An eye might stick to a knee:** Ibid.

186 **"the universe is but a watch:** Fontenelle, *Conversations*, 10.

186 **"Ah madame," he sighed:** Wright, *Franklin of Philadelphia*, 327.

186 **"You say that Beasts are Machines":** Fontenelle, *Letters of Gallantry* (London, 1715), 25 (and in a slightly different translation in Roger, *Buffon*, 118).

187 **the wrong end of his telescope:** Freedberg, *The Eye of the Lynx*, 7.

187 **"mystical letters" in those God-dictated texts:** Ruestow, *Microscope*, 54.

187 **"the Solar Atoms of light":** Wootton, *The Invention of Science*, 237, quoting Henry Power.

187 **empty chatter would "give place:** Hooke, Preface to *Micrographia*. Online at http://tinyurl.com/jq9rv8j.

187 **"discern all the secret workings:** Ibid.

188 **a "continued Chain of Ideas:** Inwood, *Forgotten Genius*, 309.

188 **"It is exceedingly difficult:** Wilson, *Invisible World*, 221.

188 **The telescope had been easier:** Ruestow, *Microscope*, 3.

188 **Even the hard-to-faze Leeuwenhoek** [FN]: Letter written on July 16, 1696, online at http://tinyurl.com/j9gf7mj.

188 **long passages of *Gulliver's Travels*:** Jonathan Swift, *Gulliver's Travels*, Part 2, Chap. 1, "A Voyage to Brobdingnag."

189 **"If our eyesight were enlarged,":** Farley, "Spontaneous Generation," 101, quoting G. de Gols, *A Theologico-Philosophical Dissertation Concerning Worms* (London, 1727).

189 **"How then can we avoid crying out:** Wilson, *Invisible World*, 190.

190 **messy, exuberant, and overflowing:** Ruestow, *Microscope*, 4.

190 **"Many a young biologist,":** Crick quoted in Vilayanur Ramachandran, "What Is Your Favorite Deep, Elegant, or Beautiful Explanation?," Edge, 2012. Online at http://tinyurl.com/jtlqzsw.

190 **Mathematics (and music and chess)** [FN]: David Eagleman's remark is from Burkhard Bilger, "The Possibilian," *New Yorker*, April 25, 2011.

190 **"it is ye perfection of God's works:** Richard Westfall, *Never at Rest: A Biography of Isaac Newton* (New York: Cambridge University Press, 1980), 327.

191 **like a suspect disappearing:** Wilson, *Invisible World*, 231. Wilson examines this topic in depth in Chapter 7, "The Microscope Superfluous and Uncertain."

191 **"legs with joints, veins in its legs:** Wilson, *Invisible World*, 190.

192 **"Why has not Man a microscopic eye?:** Alexander Pope, "An Essay on Man," Book 1.

192 **The "subjects to be enquired into:** Wilson, *Invisible World*, 226.

192 **the "great secret" would remain a secret:** Ruestow, "Images and Ideas," 211.

193 **"The history of a man:** Samuel Taylor Coleridge, *The Literary Remains of Samuel Taylor Coleridge* (London: William Pickering, 1836), 244. Online at http://tinyurl.com/jdylw8t.

194 **the new-made clock would have to tick:** Matthew Cobb makes this point in *Generation*, 226.

195 **"cardigans, symphonies, cars, and cathedrals:** Davies, *Life Unfolding*, 6.

196 **"You haven't changed a bit":** Pross, *What Is Life*, 17.

Chapter Eighteen: A Vase in Silhouette

197 **"extend it further than the astronomers:** Gasking, *Investigations*, 70.

198 **"Ye cause of gravity:** Richard Westfall, *Never at Rest: A Biography of Isaac Newton* (New York: Cambridge University Press, 1980), 505.

199 **"He loved money:** Roger, *Buffon*, xiii.

199 **"Sloths are the lowest:** Gould, "Natural History," *New York Review of Books*, Oct. 22, 1998.

199 **keep their eyes out for mammoths:** Richard Conniff, "Mammoths and Mastodons: All American Monsters," *Smithsonian*, April 2010.

199 **Buffon happily played the part:** Roger, *Buffon*, 364–365.

200 **"badly founded" and "explained nothing":** Roger, *Buffon*, 138.

200 **Buffon made a point of emphasizing:** Müller-Sievers, *Self-Generation*, 34.

200 **"in both sexes, a sort of extract:** Roger, *Buffon*, 130.

201 **a "symphony of witty and learned smut.":** J. B. Shank, *The Newton Wars and the Beginning of the French Enlightenment* (Chicago: University of Chicago Press, 2008), 435.

202 **Think of warts or tumors or moles:** See, for example, Erasmus Darwin, *Zoonomia*, 1:490.

202 **"trash, regurgitations of occultism:** Stephanson and Wagner, eds., *The Secrets of Generation*, 74, quoting Lazzaro Spallanzani.

202 **"Beware," one scientist warned:** Roe, *Matter, Life, and Generation*, 98, quoting Albrecht von Haller.

203 **"What is the difference between:** Wilson, *Invisible World*, 128.

Chapter Nineteen: Frogs in Silk Pants

209 **"undoubtedly one of the greatest:** Gasking, *Investigations*, 130.

210 **with "some apprehension":** Spallanzani, *Natural History*, 217. Online at http://tinyurl.com/zgpdteo.

210 **Spallanzani noted indignantly [FN]:** Spallanzani, *Natural History*, 200.

210 **Could it be that bats:** My description of Spallanzani's work with bats is based on the essays in *Isis* by Sven Dijkgraaf and Robert Galambos.

212 **Isaac Newton, believed in spontaneous generation:** Findlen, "Janus," 235.

212 **"So little is necessary to make an animal,":** Midgley, *Science as Salvation*, 85.

213 **Both William Harvey and Robert Hooke:** Ruestow, *Microscope*, 202, cites these examples of common beliefs and many more.

213 **"viper-powder"—dried, ground-up snake:** Thomas Birch, *The History of the Royal Society of London* (London, 1761), vol. 1. Viper powder turned up again a year later (Birch, *History*, 446, 448). Online at http://tinyurl.com/z3vb4go.

213 **"the excrements of the earth:** Thomas, *Man and the Natural World*, 55.

214 **Those ignoble animals:** Schmitt, "Spontaneous Generation," 270. See also Browne, "Noah's Flood," 109; also, Roger, *Life Sciences*, 18.

214 **"the picture of hunger":** Redi, *Vipers*, excerpt translated by M. E. Kudrati, online at http://tinyurl.com/zskm7nl.

214 **"as though it were some pearly julep":** *Redi on Vipers*, translated and annotated by Peter Knoefel, 7.

214 **"creeping up, all soft and slimy":** Redi, *Insects*, 31. Online at http://tinyurl.com/hea6j54.

215 **Presumably it was the wheat:** André Brack, ed., *The Molecular Origins of Life: Assembling Pieces of the Puzzle* (New York: Cambridge University Press, 1999), 1.

215 **The newfangled "vegetative force":** Pinto-Correia, *The Ovary of Eve*, 194.

216 **The leading view, that semen worked:** Sandler, "Re-Examination," 195–196.

217 **"in Nature, in no case:** Pinto-Correia, *The Ovary of Eve*, 198.

217 **"God creates, Linnaeus arranges,":** James Barron, "The 300th Birthday of the Man Who Organized All of Nature," *New York Times*, May 23, 2007.

217 **"the *amours*" of the frog:** Pinto-Correia, *The Ovary of Eve*, 198.

217 **"the nuptials of the newt.":** Dolman, "Spallanzani."

217 **"darting backward and forward":** Pinto-Correia, *The Ovary of Eve*, 198.

218 **"The idea of the breeches:** Meyer, *Embryology*, 174.

219 **Spallanzani owed the idea** [FN]**:** Ibid.

CHAPTER TWENTY: A DROP OF VENOM

220 **Newton conked on the head:** Newton claimed in his old age that it was the sight of an apple falling in his garden that gave him the idea that the moon, too, was falling toward the Earth. The story is controversial, and the "fact" that everyone knows—that the apple hit Newton in the head—is surely false. Richard Westfall, Newton's best biographer, discusses the falling apple in *Never at Rest: A Biography of Isaac Newton* (New York: Cambridge University Press, 1980), 154–155, and is more inclined than many to give the story some credence.

220 **Galileo lugging rocks:** Galileo supposedly performed the great drop in around 1590. David Wootton discusses the evidence pro and con in *Galileo: Watcher of the Skies* (New Haven, CT: Yale University Press, 2010), 73. Galileo certainly described such an experiment, though without ever mentioning the Leaning Tower. Historians dispute whether this was a thought experiment or an actual test.

221 **an analogy to snake venom:** Gasking, *Investigations*, 135.

221 **"priest cum scientist":** Birkhead, *Promiscuity*, 108.

222 **"resembled in color and shape:** Pinto-Correia, *The Ovary of Eve*, 207.

222 **Neither did extracts from lung or liver:** Sandler, "Re-Examination," 221.

222 **Somehow, semen was special:** Gasking, *Investigations*, 134.

222 **"I found the volume:** Ibid., 135.

223 **But when Spallanzani took:** Ibid., 136.

223 **He never put the viscous blob:** Capanna, "Spallanzani," 191.

223 **Semen samples from humans, horses:** Sandler, "Re-Examination," 208.

224 **These were bona fide animals:** Ibid., 219.

224 **"It is rather alarming:** Gasking, *Investigations*, 132.

225 **"My long experience in the world:** Metz and Monroy, eds., *Fertilization*, 18.

225 **So influential was Spallanzani:** Ibid., 10.

226 **did not "differ in the least":** Gasking, *Investigations*, 133.

226 **"Reproduction was a uniquely female occupation:** Farley, *Gametes and Spores*, 110.

CHAPTER TWENTY-ONE: THE CRAZE OF THE CENTURY

228 **in a Paris courtyard:** This account is from Pera, *Ambiguous Frog*, 16–19.

229 **"the craze of the century.":** Whitaker, Smith, and Finger, eds., *Brain, Mind and Medicine*, 271.

231 **Audiences wept with Lear:** William Shakespeare, *King Lear*, Act 5, Scene 3.

232 **was in fact "electrical fluid":** Gasking, *Experimental Biology*, 104.

232 **endless opportunities for creative mayhem:** Steven Johnson highlights this contrast between gravity and electricity in *The Invention of Air: A Story of Science, Faith, Revolution, and the Birth of America* (New York: Riverhead, 2009), 20.

232 **Traveling lecturers set off foot-long sparks:** Hochadel, "Shock," 55–56.

232 **"electricity took place of quadrille.":** "Experiments on Electricity," *Gentleman's Magazine* 15 (1745): 194. Online at http://tinyurl.com/jun7uq4.

232 **don a pair of glass slippers:** Ashcroft, *Spark of Life*, 16.

233 **Electricity was always described:** Pera, *Ambiguous Frog*, 3–5.

233 **In England, France, Italy, even Poland:** Bensaude-Vincent and Blondel, eds., *Science and Spectacle*, 75, and "Experiments on Electricity."

233 **"honored [the local electrical savants]:** "Experiments on Electricity."

233 **"It is singular to see:** Heilbron, *Electricity*, 318.

234 **1,800 tingling, tormented soldiers:** Ashcroft, *Spark of Life*, 15.

234 **"I inadvertently took the whole [shock]:** Brox, *Brilliant*, 98, and the American Physical Society, "Ben Franklin," online at http://tinyurl.com/htzuxmn.

234 **it demonstrated that the lightning bolt:** Brox, *Brilliant*, 100, and Krider, "Benjamin Franklin."

234 **"heaven's artillery.":** Stacy Schiff, *The Witches: Salem, 1692* (New York: Little, Brown, 2015), 17 and 427n.

234 **calamitous to the bell ringers:** Brox, *Brilliant*, 99, and Cohen, *Franklin's Science*, especially the section entitled "Lightning Rods Versus Church Bells," beginning on 119.

234 **Many religious believers opposed [FN]:** Cohen, *Franklin's Science*, 159.

235 **The first person to die:** "Account of the Death of Georg Richmann," *Pennsylvania Gazette*, March 5, 1754. Online at http://tinyurl.com/zrm7ja4.

235 **Joseph Priestley** [FN]: Cohen, *Franklin's Science*, 6.

235 **"The first time I experienced it,":** Heilbron, *Electricity*, 18.

235 **left her temporarily unable to walk:** Heilbron, *Electricity*, 314, and Ashcroft, *Spark of Life*, 14.

235 **"Instead of one of the two hands":** Strickland, "Ideology of Self-Knowledge," 458.

236 **where he met Benjamin Franklin:** Porter, *The Facts of Life*, 108.

236 **In London Graham opened a Temple:** Syson, *Doctor of Love*, 3.

236 **"the exhilarating force of electrical fire":** Porter, *The Facts of Life*, 109.

236 **Graham promised "any gentleman:** Ibid., 110.

237 **"Even the venereal act itself":** Otto, "Regeneration of the Body," and Stephanson and Wagner, eds., *The Secrets of Generation*, 14.

CHAPTER TWENTY-TWO:
"I SAW THE DULL YELLOW EYE OF THE CREATURE OPEN"

238 **Luigi Galvani's claim:** See Piccolino and Bresadola, *Shocking Frogs*, 1–25; also George Johnson, *The Ten Most Beautiful Experiments*, 60–74, and Pera, *Ambiguous Frog*, xx–xxvi.

239 **a frog broth:** Historians have repeated the tale since Jean-Louis Alibert recorded it in his *Eloge Historique de Galvani*, published in 1801. For a skeptical modern take, see Piccolino and Bresadola, *Shocking Frogs*, 5.

239 **by the French Revolution:** Nicholas Wade, Marco Piccolino, and Adrian Simmons, "Luigi Galvani." Online at http://tinyurl.com/hlogn33.

239 **"a genius for electricity.":** Pera, *Ambiguous Frog*, 41.

240 **"He had invented the first:** Ashcroft, *Spark of Life*, 25.

240 **The world "laughs at me,":** The quote is firmly entrenched in scientific legend. Frances Ashcroft, herself an esteemed chemist, quotes the remark in *Spark of Life* (8), but she comments in her notes that it was likely invented by a French astronomer, Camille Flammarion, in 1862.

241 **"the first of the decapitated criminals:** Ashcroft, *Spark of Life*, 28.

242 **Spectators gasped that Foster had come back:** "George Foster," *Newgate Calendar*, Jan. 1803. Online at http://tinyurl.com/hpjcpav.

242 **Spectators vomited and fainted:** Ashcroft, *Spark of Life*, 29.

243 **June 1816, near Lake Geneva:** My account of that fateful summer is based largely on Johnson, "Mary Shelley and Her Circle," in *A Life with Mary Shelley*, and Stott, *The Poet and the Vampyre*.

243 **"We were placed hand in hand:** Darby Lewes, ed., *A Brighter Morn: The Shelley Circle's Utopian Project* (Lanham, MD: Lexington Books, 2002), 147.

244 **At Oxford, Shelley's room overflowed:** Richard Holmes, *Shelley: The Pursuit* (New York: Harper, 2005), 37.

244 **no more than electrified clay":** Ashcroft, *Spark of Life*, 9.

244 **He grew so fat:** Jenny Uglow, *The Lunar Men: Five Friends Whose Curiosity Changed the World* (New York: Farrar, Straus and Giroux, 2002), xiv.

245 **"preserved a piece of vermicelli:** Mary Shelley, *Frankenstein*, Introduction to the 1831 edition. Online at http://tinyurl.com/zdse7rx. See also Ashton Nichols, "Erasmus Darwin and the Frankenstein 'Mistake,'" Romantic Natural History. Online at http://tinyurl.com/jsc34ho.

245 **"Perhaps a corpse would be re-animated":** Shelley, *Frankenstein*, Introduction to the 1831 edition.

245 **"It was on a dreary night of November:** Mary Shelley, *Frankenstein*, Chap. 5. Online at http://tinyurl.com/hpuarvv.

CHAPTER TWENTY-THREE: THE NOSE OF THE SPHINX

247 **"Only about one millionth of one billionth:** Alan Lightman, "Our Lonely Home in Nature," *New York Times*, May 6, 2014.

247 **Robert Brown focused instead:** The source for this section on Robert Brown is Nott, "Molecular Reality."

249 **The term "organic chemistry":** Brian Silver, *The Ascent of Science* (New York: Oxford University Press, 2000), 319.

250 **"I can make urea:** Ramberg, "The Death of Vitalism," 178.

250 **Lavoisier's discoveries** [FN]**:** For a (skeptical) examination of the quote attributed to the revolutionary judge, see Dennis I. Duveen, "Antoine Laurent Lavoisier and the French Revolution," *Journal of Chemical Education* 31 (February 1954).

250 **In one painstaking experiment:** Hoffmann, *Life's Ratchet*, 30.

251 **But Helmholtz's new experiments:** Ibid., 41.

251 **F. Scott Fitzgerald famously suggested:** Fitzgerald, *The Crack-Up* (New York: New Directions, 2009), 69.

CHAPTER TWENTY-FOUR: "THE GAME IS AFOOT"

252 **Across Europe, scientists had responded:** Gasking, *Investigations*, 138–139.

253 **"surpasses the utmost powers:** Peter Roget, *The Bridgewater Treatises on the Power Wisdom and Goodness of God As Manifested in the Creation* (London: William Pickering, 1834), 2:582. Online at http://tinyurl.com/glwua06.

253 **A decade before, two young colleagues:** Gasking, *Investigations*, 140–142.

254 **What modern thinker wanted to return:** Farley, *Gametes and Spores*, 47.

254 **"take pleasure in boring us:** La Mettrie, *Machine Man*, 86.

255 **"There will never be a Newton:** Immanuel Kant, *Critique of Judgment* (London: Macmillan, 1914), 312. Kant's remark is often cited in the form I quote in the text; J. H. Bernard's translation here differs slightly. Online at http://tinyurl.com/jc9fjbm.

256 **"When I observed the ovary,":** Baer, "On the Genesis of the Ovum," 120.

256 **"Every animal which is begotten:** Baer and Sarton, "Discovery," 324.

257 **a fluid that "curdled":** Ibid., 317.

257 **On the last day:** Ibid., 325.

CHAPTER TWENTY-FIVE: CAUGHT!

258 **On an October evening:** Otis, *Müller's Lab*, 63.

259 **The two men left their coffee:** Vasil, "History of Plant Biotechnology," 1424, and Bechtel, *Discovering Cell Mechanisms*, 68–69.

259 **"All cells come from cells.":** Wagner, "Virchow," 917.

259 **"all developed tissues can be traced back:** Virchow, *Cellular Pathology* (London: John Churchill, 1860), 27.

259 **a host of colleagues and rivals:** Hunter, *Vital Forces*, 64–74.

260 **"a great many little Boxes":** Hooke, *Micrographia*, Observation 18, "Of the Schematisme or Texture of Cork." Online at http://tinyurl.com/gwhkl5s.

260 **All through the 1700s:** Hunter, in *Vital Forces*, describes a series of technical improvements in microscope design that dated from the 1830s (see 60–61).

260 **Botanists spent their time:** Gasking, *Investigations*, 168.

260 **"stark fools and idiots":** Rudolf Virchow, *The Freedom of Science in the Modern State* (London: John Murray, 1878), 18.

260 **the thrilling discovery of a Neanderthal skull:** Schultz, "Rudolf Virchow."

261 **"civil war between cells.":** Wagner, "Virchow," 918.

261 **He chose sausages:** Schultz, "Rudolf Virchow." But though the story is a favorite of medical historians, it may well be apocryphal. It may have arisen from a true event, in 1865, when Virchow delivered a speech in Berlin calling for the inspection of pork as a public health precaution. Virchow held up a smoked sausage for the crowd to see and explained that this delectable treat was in fact infected. A veterinarian in the audience stood up and declared that trichinae, the organisms that cause trichinosis, were "the most harmless animals in the world. It is only doctors without practice who make a noise about them, in order to create some occupation for themselves." One of Virchow's colleagues challenged the man to eat the infected sausage. The crowd began chanting, "Eat! Eat!" The veterinarian gave in, took a bite, and stormed out of the hall. Five days later, according to newspaper reports, he lay helpless in bed, unable to move arms or legs. (This account is from Dr. Thudichum, "The Trichina Disease," *Edinburgh Medical Journal* 11, part 2 [1866]: 771–772. Online at http://tinyurl.com/glya6ps.)

261 **Living organisms followed "blind laws:** Hunter, *Vital Forces*, 63.

261 **Hertwig was frosty and forbidding:** Goldschmidt, *Portraits*, 76–80.

262 **"arises to completion like a sun:** Weindling, *Darwinism*, 70.

263 **Driesch was snobbish:** Goldschmidt, *Portraits*, 69.

264 **Life, he declared, contained mysteries:** Ibid.

264 **The resolution of the mystery:** Fisher, *Weighing the Soul*, 138–140.

265 **the writer John Stewart Collis:** Collis, *The Worm Forgives the Plough*, 43.

BIBLIOGRAPHY

Anstey, Peter R. "Boyle on Seminal Principles." *Studies in History and Philosophy of Biological and Biomedical Sciences* 33 (2002).

Aristotle. *On the Generation of Animals.* Online at http://tinyurl.com/jsjqwmp.

Ashcroft, Frances. *The Spark of Life: Electricity in the Human Body.* New York: Norton, 2013.

Aubrey, John. *Brief Lives.* Woodbridge, UK: Boydell, 1982.

Aughterson, Kate, ed. *Renaissance Woman: A Sourcebook: Constructions of Femininity in England.* New York: Routledge, 1995.

Augustine, *The City of God.* Online at http://tinyurl.com/gpsx7la.

Baer, Karl Ernst von. "On the Genesis of the Ovum of Mammals and of Man." *Isis* 47, no. 2 (June 1956).

Baer, Karl Ernst von, and George Sarton. "The Discovery of the Mammalian Egg and the Foundation of Modern Embryology." *Isis* 16, no. 2 (Nov. 1931).

Bainbridge, David. *Making Babies: The Science of Pregnancy.* Cambridge, MA: Harvard University Press, 2001.

Bates, A. W. *The Anatomy of Robert Knox: Murder, Mad Science and Medical Regulation in Nineteenth Century Edinburgh.* Eastbourne, UK: Sussex, 2010.

Bechtel, William. *Discovering Cell Mechanisms: The Creation of Modern Cell Biology.* New York: Cambridge University Press, 2006.

Beckerman, S., and P. Valentine, eds. *Cultures of Multiple Fathers: The Theory and Practice of Partible Paternity in Lowland South America.* Gainesville: University Press of Florida, 2002.

Bensaude-Vincent, Bernadette, and Christine Blondel, eds. *Science and Spectacle in the European Enlightenment.* Aldershot, UK: Ashgate, 2007.

Birkhead, Tim. *Promiscuity: An Evolutionary History of Sperm Competition.* Cambridge, MA: Harvard University Press, 2000.

———. *The Wisdom of Birds: An Illustrated History of Ornithology.* New York: Bloomsbury, 2008.

Boyce, James. *Born Bad: Original Sin and the Making of the Western World.* Berkeley: Counterpoint, 2015.

Browne, Janet. "Noah's Flood, the Ark, and the Shaping of Early Modern Natural History." Chapter 5 in *When Science and Christianity Meet*, ed. David C. Lindberg and Ronald L. Numbers. Chicago: University of Chicago Press, 2003.

Brox, Jane. *Brilliant: The Evolution of Artificial Light*. New York: Houghton Mifflin, 2010.

Bynum, Caroline Walker. *Fragmentation and Redemption: Essays on Gender and the Human Body in Medieval Religion*. New York: Zone, 1992.

Capanna, Ernesto. "Lazzaro Spallanzani: At the Roots of Modern Biology." *Journal of Experimental Zoology* 285 (1999).

Clayton, Martin, and Ronald Philo. *Leonardo da Vinci: The Mechanics of Man*. Los Angeles: Getty Publications, 2010.

Cobb, Matthew. *Generation: The Seventeenth-Century Scientists Who Unraveled the Secrets of Sex, Life, and Growth*. New York: Bloomsbury, 2006.

Cody, Lisa Forman. *Birthing the Nation: Sex, Science, and the Conception of Eighteenth-Century Britons*. New York: Oxford University Press, 2008.

Cohen, I. Bernard. *Benjamin Franklin's Science*. Cambridge, MA: Harvard University Press, 1990.

Collis, John Stewart. *The Worm Forgives the Plough*. Pleasantville, NY: Akadine, 1997.

Cutler, Alan. *The Seashell on the Mountaintop: How Nicolaus Steno Solved an Ancient Mystery and Created a Science of the Earth*. New York: Penguin, 2004.

Darwin, Erasmus. *Zoonomia*, vol. 1. London, 1794.

Davies, Jamie A. *Life Unfolding: How the Human Body Creates Itself*. Oxford: Oxford University Press, 2014.

Dawson, Virginia P. *Nature's Enigma: The Problem of the Polyp in the Letters of Bonnet, Trembley, and Réaumur*. Philadelphia: American Philosophical Society, 1987.

———. "Regeneration, Parthenogenesis, and the Immutable Order of Nature." *Archives of Natural History* 18, no. 3 (1991).

De Graaf, Regnier. *Regnier de Graaf on the Human Reproductive Organs*. Translated by H. D. Jocelyn and B. P. Setchell. Oxford: Blackwell, 1972.

Diamond, Jared. *Why Is Sex Fun? The Evolution of Human Sexuality*. New York: Basic Books, 1997.

Dijkgraaf, Sven. "Spallanzani's Unpublished Experiments on the Sensory Basis of Object Perception in Bats." *Isis* 51, no. 1 (March 1960).

Dobell, Clifford. *Antony van Leeuwenhoek and His "Little Animals."* New York: Russell and Russell, 1958.

Dolman, Claude E. "Lazzaro Spallanzani." *The Complete Dictionary of Scientific Biography*, encyclopedia.com, 2008.

Farley, John. *Gametes and Spores: Ideas About Sexual Reproduction 1750–1914*. Baltimore: Johns Hopkins University Press, 1982.

———. "The Spontaneous Generation Controversy (1700–1860): The Origin of Parasitic Worms." *Journal of the History of Biology* 5, no. 1 (Spring 1972).

Findlen, Paula. "The Janus Faces of Science in the Seventeenth Century: Athanasius Kircher and Isaac Newton." In *Rethinking the Scientific Revolution*, ed. Margaret J. Osler. New York: Cambridge University Press, 2000.

Fisher, Len. *Weighing the Soul: Scientific Discovery from the Brilliant to the Bizarre*. New York: Arcade, 2004.

Flandrin, Jean Louis. *Sex in the Western World: The Development of Attitudes and Behavior.* Chur, Switzerland: Harwood, 1991.

Fletcher, Anthony. *Gender, Sex, and Subordination in England, 1500–1800.* New Haven, CT: Yale University Press, 1995.

Fontenelle, Bernard de. *Conversations on the Plurality of Worlds.* London, 1803.

———. *Letters of Gallantry.* London, 1715.

Freedberg, David. *The Eye of the Lynx: Galileo, His Friends, and the Beginning of Modern Natural History.* Chicago: University of Chicago Press, 2002.

Fudge, Erica, ed. *Renaissance Beasts: Of Animals, Humans, and Other Wonderful Creatures.* Urbana: University of Illinois Press, 2004.

Galambos, Robert. "The Avoidance of Obstacles by Flying Bats: Spallanzani's Ideas (1794) and Later Theories." *Isis* 34, no. 2 (Autumn 1942).

Gasking, Elizabeth. *Investigations into Generation, 1651–1828.* Baltimore, MD: Johns Hopkins University Press, 1967.

———. *The Rise of Experimental Biology.* New York: Random House, 1970.

———. "The Spontaneous Generation Controversy (1700–1860): The Origin of Parasitic Worms." *Journal of the History of Biology* 5, no. 1 (Spring 1972).

Goldschmidt, Richard B. *Portraits from Memory: Recollections of a Zoologist.* Seattle: University of Washington Press, 1956.

Gonzalez-Crussi. *Carrying the Heart: Exploring the Worlds Within Us.* New York: Kaplan, 2009.

Gould, Stephen Jay. "Adam's Navel." In *The Flamingo's Smile: Reflections in Natural History.* New York: Norton, 1987.

———. "The Man Who Invented Natural History." *New York Review of Books*, Oct 22, 1998.

Gregory, Mary Efrosini. *Evolutionism in Eighteenth-Century French Thought.* New York: Peter Lang, 2008.

Gross, Charles G. "Rembrandt's 'The Anatomy Lesson of Dr. Joan Deijman.'" *Trends in Neuroscience* 21, no. 6 (June 1, 1998).

Harrison, Peter. "Reading Vital Signs: Animals and the Experimental Philosophy." In *Renaissance Beasts: Of Animals, Humans, and Other Wonderful Creatures*, ed. Erica Fudge. Urbana: University of Illinois Press, 2004.

Heckscher, William S. *Rembrandt's Anatomy of Dr. Nicholaas Tulp: An Iconological Study.* New York: New York University Press, 1958. Online at http://archive.org /stream/rembrandt00heck/rembrandt00heck_djvu.txt.

Heilbron, J. L. *Electricity in the 17th and 18th Centuries: A Study of Early Modern Physics.* Berkeley: University of California Press, 1979.

Hiatt, Lester Richard. *Arguments About Aborigines: Australia and the Evolution of Social Anthropology.* New York: Cambridge University Press, 1996.

Hochadel, Oliver. "A Shock to the Public: Itinerant Lecturers and Instrument Makers as Practitioners of Electricity in the German Enlightenment (1740–1800)." Online at http://tinyurl.com/zczaqe9.

Hoffmann, Peter M. *Life's Ratchet: How Molecular Machines Extract Order from Chaos*. New York: Basic Books, 2012.

Home, R. W. "Force, Electricity, and the Powers of Living Matter in Newton's Mature Philosophy of Nature." In *Religion, Science, and Worldview: Essays in Honor of Richard Westfall*, ed. Margaret J. Osler and Paul Lawrence Farber. New York: Cambridge University Press, 1985.

Hunter, Graeme K. *Vital Forces: The Discovery of the Molecular Basis of Life*. San Diego: Academic Press, 2000.

Inhorn, Marcia. *Quest for Conception: Gender, Infertility, and Egyptian Medical Traditions*. Philadelphia: University of Pennsylvania Press, 1994.

Inwood, Stephen. *The Forgotten Genius: The Biography of Robert Hooke, 1635–1703*. San Francisco: MacAdam/Cage, 2005.

Jacobs, Alan. *Original Sin: A Cultural History*. New York: HarperOne, 2008.

Jacquart, Danielle, and Claude Thomasset. *Sexuality and Medicine in the Middle Ages*. Cambridge: Polity, 1998.

Jardine, Lisa. *The Curious Life of Robert Hooke: The Man Who Measured London*. New York: Harper, 2004.

Johnson, Barbara. *A Life with Mary Shelley*. Stanford, CA: Stanford University Press, 2014.

Johnson, George. *The Ten Most Beautiful Experiments*. New York: Knopf, 2008.

Jones, Jonathan. *The Lost Battles: Leonardo, Michelangelo, and the Artistic Duel That Defined the Renaissance*. New York: Knopf, 2012.

Jorink, Eric. "Between Emblematics and the 'Argument from Design': The Representation of Insects in the Dutch Republic." In *Early Modern Zoology: The Construction of Animals in Science, Literature, and the Visual Arts*, ed. Karl A. E. Enenkel and Paul J. Smith. Leiden, Netherlands: Brill, 1997.

Keele, Kenneth. *Leonardo da Vinci: Anatomical Drawings from the Royal Library Windsor Castle*. New York: Metropolitan Museum of Art, 1983.

———. *Leonardo da Vinci's Elements of the Science of Man*. New York: Academic Press, 2014.

Kemp, Martin, ed. *Leonardo on Painting: An Anthology of Writings by Leonardo da Vinci*. New Haven, CT: Yale University Press, 1991.

Keynes, Geoffrey. *The Life of William Harvey*. Oxford: Oxford University Press, 1966.

King-Hele, Desmond. *Erasmus Darwin and the Romantic Poets*. New York: Macmillan, 1986.

Kottek, Samuel S. "Embryology in Talmudic and Midrash Literature." *Journal of the History of Biology* 14, no. 2 (Fall 1981).

Krider, E. Philip. "Benjamin Franklin and Lightning Rods." *Physics Today* 59, no. 1 (2006). Online at http://tinyurl.com/gwgs3ln.

La Mettrie, Julien de. *Machine Man and Other Writings*. New York: Cambridge University Press, 1996.

Laqueur, Thomas. *Making Sex: Body and Gender from the Greeks to Freud*. Cambridge, MA: Harvard University Press, 1990.

———. *Solitary Sex: A Cultural History of Masturbation*. New York: Zone, 2004.

Leroi, Armand Marie. *The Lagoon: How Aristotle Invented Science*. New York: Viking, 2014.

Malinowski, Bronislaw. "Baloma; The Spirits of the Dead in the Trobriand Islands." *Journal of the Royal Anthropological Institute of Great Britain and Ireland* 46 (1916). Online at http://tinyurl.com/z2vxxuq.

Manniche, Lisa. *Sexual Life in Ancient Egypt*. London: Routledge, 2004.

Maupertuis, Pierre-Louis de. *The Earthly Venus*. New York: Johnson Reprint, 1966.

McLaren, Angus. *Reproductive Rituals: The Perception of Fertility in England from the Sixteenth Century to the Nineteenth Century*. New York: Methuen, 1984.

McMahon, Darrin. *Divine Fury: A History of Genius*. New York: Basic Books, 2013.

Merchant, Carolyn. *The Death of Nature: Women, Ecology and the Scientific Revolution*. New York: Harper, 1989.

Metz, Charles B., and Alberto Monroy, eds. *Fertilization: Comparative Morphology, Biochemistry, and Immunology*. New York: Academic Press, 1967.

Meyer, Arthur William. *The Rise of Embryology*. Stanford, CA: Stanford University Press, 1939.

Midgley, Mary. *Science as Salvation: A Modern Myth and Its Meaning*. New York: Routledge, 1992.

Miller, Jonathan. *The Body in Question*. New York: Random House, 1978.

Montillo, Roseanne. *The Lady and Her Monsters: A Tale of Dissections, Real-Life Dr. Frankensteins, and the Creation of Mary Shelley's Masterpiece*. New York: William Morrow, 2013.

Moore, John A. *Science as a Way of Knowing: The Foundations of Modern Biology*. Cambridge, MA: Harvard University Press, 1993.

Müller-Sievers, Helmut. *Self-Generation: Biology, Philosophy, and Literature Around 1800*. Stanford, CA: Stanford University Press, 1997.

Needham, Joseph. *A History of Embryology*. Cambridge: Cambridge University Press, 1959.

Nichols, Ashton. "Erasmus Darwin and the Frankenstein 'Mistake.'" Online at http://tinyurl.com/jsc34ho.

Nightingale, Andrea. *Once Out of Nature: Augustine on Time and the Body*. Chicago: University of Chicago Press, 2011.

Nott, Mick. "Molecular Reality: The Contributions of Brown, Einstein and Perrin." *School Science Review* 86, no. 317 (June 2005).

Otis, Laura. *Müller's Lab: The Story of Jakob Henle, Theodor Schwann, Emil du Bois-Reymond, Hermann von Helmholtz, Rudolf Virchow, Robert Remak, Ernst Haeckel, and Their Brilliant, Tormented Advisor*. New York: Oxford University Press, 2007.

Otto, Peter. "The Regeneration of the Body: Sex, Religion and the Sublime in James Graham's *Temple of Health and Hymen*." *Romanticism and Sexuality*, no. 23 (August 2001). Online at http://tinyurl.com/zl52yxs.

Pantel, Pauline Schmitt, ed. *A History of Women in the West, Volume I: From*

Ancient Goddesses to Christian Saints. Cambridge, MA: Harvard University Press, 1994.

Park, Katharine. "The Criminal and the Saintly Body: Autopsy and Dissection in Renaissance Italy." *Renaissance Quarterly* 47, no. 1 (Spring 1994).

————. "Dissecting the Female Body: From Women's Nature to the Secrets of Bodies." In *Attending to Early Modern Women*, ed. Adele Seeff and Jane Donawerth. Newark: University of Delaware Press, 2000.

————. *Secrets of Women: Gender, Generation, and the Origins of Human Dissection.* New York: Zone, 2010.

Pera, Marcello. *The Ambiguous Frog: The Galvani-Volta Controversy on Animal Electricity.* Princeton, NJ: Princeton University Press, 1992.

Piccolino, Marco, and Marco Bresadola. *Shocking Frogs: Galvani, Volta, and the Electric Origins of Neuroscience.* New York: Oxford University Press, 2013.

Pinto-Correia, Clara. *The Ovary of Eve: Egg and Sperm and Preformation.* Chicago: University of Chicago Press, 1997.

Porter, Roy. *The Facts of Life: The Creation of Sexual Knowledge in Britain, 1650–1950.* New Haven, CT: Yale University Press, 1995.

Power, D'Arcy. *William Harvey.* London: Fisher Unwin, 1897.

Pross, Addy. *What Is Life? How Chemistry Becomes Biology.* Oxford, UK: Oxford University Press, 2012.

Ramberg, Peter J. "The Death of Vitalism and the Birth of Organic Chemistry: Wohler's Urea Synthesis and the Disciplinary Identity of Organic Chemistry." *Ambix* 47, no. 3 (Nov. 2000).

Redi, Francesco. *Experiments on the Generation of Insects.* Chicago: Open Court, 1909. Online at http://tinyurl.com/hea6j54.

————. *Francesco Redi on Vipers.* Translated and annotated by Peter Knoefel. Leiden, Netherlands: Brill, 1988.

Richardson, Ruth. *Death, Dissection and the Destitute.* Chicago: University of Chicago Press, 2001.

Riddle, John M. *Contraception and Abortion from the Ancient World to the Renaissance.* Cambridge, MA: Harvard University Press, 1994.

Riskin, Jessica. "The Defecating Duck, or, the Ambiguous Origins of Artificial Life." *Critical Inquiry* 29, no. 4 (Summer 2003).

Roe, Shirley A. *Matter, Life, and Generation: Eighteenth-Century Embryology and the Haller-Wolff Debate.* New York: Cambridge University Press, 1981.

Roger, Jacques. *Buffon: A Life in Natural History.* Ithaca, NY: Cornell University Press, 1997.

————. *The Life Sciences in Eighteenth-Century French Thought.* Stanford, CA: Stanford University Press, 1997.

————. "Two Scientific Discoveries: Their Genesis and Destiny." In *On Scientific Discovery: The Erice Lectures 1977*, ed. Mirko Grmek, Robert Cohen, and Guido Cimino. Dordrecht, Netherlands: Reidel, 1977.

Rosner, Lisa. *The Anatomy Murders: Being the True and Spectacular History of*

Edinburgh's Notorious Burke and Hare and the Man of Science Who Abetted Them in the Commission of Their Most Heinous Crimes. Philadelphia: University of Pennsylvania Press, 2009.

Ruestow, Edward G. "Images and Ideas: Leeuwenhoek's Perception of the Spermatozoa." *Journal of the History of Biology* 16, no. 2 (Summer 1983).

———. "Leeuwenhoek and the Campaign Against Spontaneous Generation." *Journal of the History of Biology* 17, no. 2 (Summer 1984).

———. *The Microscope in the Dutch Republic: The Shaping of Discovery* (New York: Cambridge University Press, 1996).

———. "Piety and the Defense of Natural Order: Swammerdam on Generation." In *Religion, Science, and Worldview: Essays in Honor of Richard Westfall*, ed. Margaret J. Osler and Paul Lawrence Farber. New York: Cambridge University Press, 1985.

Sandler, Iris. "The Re-Examination of Spallanzani's Interpretation of the Role of the Spermatic Animalcules in Fertilization." *Journal of the History of Biology* 6, no. 2 (Fall 1973).

Schmitt, William J. "Spontaneous Generation and Creation." *Thought* 37, no. 2 (1962).

Schultz, Myron. "Rudolf Virchow." *Emerging Infectious Diseases* 14, no. 9 (Sept. 2008). Online at http://tinyurl.com/ma9bpk3.

Shelley, Mary Wollstonecraft. *Frankenstein*. 1818/1831. Online at http://tinyurl.com/jlelu2x.

Shorter, Edward. *A History of Women's Bodies*. New York: Basic Books, 1982.

Sleigh, Charlotte. "Jan Swammerdam's Frogs." *Notes and Records of the Royal Society* 66 (2012).

Smith, Justin. *Divine Machines: Leibniz and the Sciences of Life*. Princeton, NJ: Princeton University Press, 2011.

———. "Leibniz on Spermatozoa and Immortality." *Archiv für Geschichte der Philosophie* 89, no. 3 (2007).

Snyder, Laura J. *Eye of the Beholder: Johannes Vermeer, Antoni van Leeuwenhoek, and the Reinvention of Seeing*. New York: Norton, 2015.

Spallanzani, Lazzaro. *Dissertations Relative to the Natural History of Animals and Vegetables*. London, 1784. Online at http://tinyurl.com/zgpdteo.

Stephanson, Raymond, and Darren Wagner, eds. *The Secrets of Generation: Reproduction in the Long Eighteenth Century*. Toronto: University of Toronto Press, 2015.

Stott, Andrew McConnell. *The Poet and the Vampyre: The Curse of Byron and the Birth of Literature's Greatest Monsters*. New York: Pegasus Books, 2014.

Stott, Rebecca. *Darwin's Ghosts: The Secret History of Evolution*. New York: Spiegel & Grau, 2012.

Strickland, Stuart Walker. "The Ideology of Self-Knowledge and the Practice of Self-Experimentation." *Eighteenth-Century Studies* 31, no. 4 (Summer 1998).

Suh, H. Anna, ed. *Leonardo's Notebooks: Writing and Art of the Great Master*. New York: Black Dog & Leventhal, 2005.

Syson, Lydia. *Doctor of Love: James Graham and His Celestial Bed*. London: Alma, 2012.

Terrall, Mary. *The Man Who Flattened the Earth: Maupertuis and the Sciences in the Enlightenment*. Chicago: University of Chicago Press, 2002.

Thomas, Keith. *Man and the Natural World*. New York: Pantheon, 1983.

Tissot, Samuel. *A Treatise on the Diseases Produced by Onanism*. New York: Collins and Hannay, 1832.

Todd, Dennis. *Imagining Monsters: Miscreations of the Self in Eighteenth-Century England*. Chicago: University of Chicago Press, 2005.

Trounson, Alan O., and Roger G. Gosden, eds. *Biology and Pathology of the Oocyte: Its Role in Fertility and Reproductive Medicine*. New York: Cambridge University Press, 2003.

Vartanian, Aram. "Trembley's Polyp, La Mettrie, and Eighteenth-Century French Materialism." *Journal of the History of Ideas* 11, no. 3 (June 1950).

Vasil, Indra K. "A History of Plant Biotechnology: From the Cell Theory of Schleiden and Schwann to Biotech Crops." *Plant Cell Reports* 27 (2008).

Vesalius, Andreas. *On the Fabric of the Human Body, Book I: The Bones and Cartilages*. Translated by William Frank Richardson. San Francisco: Norman, 1998.

———. *On the Fabric of the Human Body, Book V: The Organs of Nutrition and Generation*. Translated by William Frank Richardson. Novato, CA: Norman, 2007.

Virchow, Rudolf. *Cellular Pathology*. London: John Churchill, 1860.

Wagner, Robert P. "Rudolf Virchow and the Somatic Basis of Somatic Ecology." *Genetics* 151 (March 1999).

Weindling, Paul. *Darwinism and Social Darwinism in Imperial Germany: The Contribution of the Cell Biologist Oscar Hertwig*. Stuttgart, Germany: Gustav Fischer, 1991.

Westfall, Richard S. *Science and Religion in Seventeenth-Century England*. Ann Arbor: University of Michigan Press, 1973.

Whitaker, Harry, C. U. M. Smith, and Stanley Finger, eds. *Brain, Mind and Medicine: Essays in Eighteenth-Century Neuroscience*. New York: Springer, 2007.

Wiesner, Merry. *Women and Gender in Early Modern Europe*. New York: Cambridge University Press, 2000.

Wilson, Catherine. *The Invisible World: Early Modern Philosophy and the Invention of the Microscope*. Princeton, NJ: Princeton University Press, 1995.

Wootton, David. *The Invention of Science: A New History of the Scientific Revolution*. New York: Harper, 2015.

Wright, Esmond. *Franklin of Philadelphia*. Cambridge, MA: Harvard University Press, 1986.

Wright, Thomas. *William Harvey: A Life in Circulation*. New York: Oxford University Press, 2012.

Zimmer, Carl. *Soul Made Flesh: The Discovery of the Brain—and How It Changed the World*. New York: Free Press, 2004.

INDEX

LYNN GOLDEN

EDWARD DOLNICK is the former chief science writer for the *Boston Globe* and the author of *The Clockwork Universe*, *The Forger's Spell*, *Down the Great Unknown*, *The Rush*, *Madness on the Couch*, and the Edgar Award–winning *The Rescue Artist*. He splits his time between Virginia and New York City.